Science
and
Key of Life

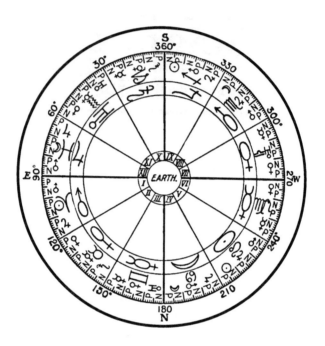

Science
and
Key of Life

PLANETARY INFLUENCES

VOLUME 6 OF 7

Alvidas

SAMUEL WEISER, INC.

York Beach, Maine

Published in 1997 by
Samuel Weiser, Inc.
P. O. Box 612
York Beach, ME 03910-0612

Library of Congress Cataloging-in-Publication Data

Alvidas.
    Science and key of life : planetary influences / Alvidas.
      p.   cm.
    Originally published: Detroit : Astro Pub. Co., 1902.
    ISBN 0-87728-911-5 (set : alk. paper)
    1. Astrology. I. Title.
BF1708.1.A44   1997
133.5--dc21
                            97-367
                            CIP

Volume 6: ISBN 0-87728-917-4
7 Volume set: ISBN 0-87728-911-5
BJ

Printed in the United States of America

The paper used in this publication meets the minimum
requirements of the American National Standard for
Permanence of Paper for Printed Library
Materials Z39.48-1984.

CONTENTS

CHAPTER XVII.

CHAPTER XVIII.

CHAPTER XIX.

CHAPTER XX.

CHAPTER XXI.

CHAPTER XXII.

CHAPTER XXIII.

CHAPTER XXIV.

PREFACE

In adding this volume to the five already published of this series—Science and Key of Life, Planetary Influences—the undersigned is imbued with the impelling thought that it contains added truth of incalculable value in the solution of the problems of human existence: truths that will answer in part the great inquiry, from whence, whither and what are the purposes of life upon this Earth planet.

Believing that the time has come when the science of stellar influence must be recognized as bearing its all important part in bringing humanity to a plane of broader consciousness and higher attainment, this volume and the seventh which will immediately follow, are published. While they continue the treatment of subjects brought to view in the preceding volumes, they also contain much that is marvelous regarding man and his relation to infinite law: of which the world is largely ignorant.

It is in the fulfillment of a sense of sacred obligation to his fellow men that these closing volumes of this work are presented to the minds of thoughtful people as a help in working out their mission by the exercise of free will actuated and prompted by the all-powerful God-like spirit of love and kindness.

For ages the destiny of the world, aye, the universe, has been mapped out in the sky, but there have been few of mankind pure enough to interpret the handwriting of the Supreme Ruler. We may only realize through pain and sorrow, that we alone are the builders of our own future and the rulers of our own destiny.

Grand are the symbols of being, but that which is symbolized is
    greater.
Vast the created and beheld, but vaster the inward creator.
Back of the sound broods the silence, back of the gift stands the
    giving;
Back of the hand that receives, are the thrice sensitive nerves of
    receiving.

CHAPTER I

The Sacredness of Life.

There is a wonderful thought to consider in association with the desire for progress on the part of humanity, and how great a part this desire for success plays in the physical manifestations of mankind to-day. This is clearly evident from the general trend of inquiries expressed by men and women, and apart from any special advice applicable to each individual case it will be of benefit to consider the matter in a broad light, as, for instance, what are the most essential conditions relating to the progress and success of the human race to-day.

In the first place, it must be recognized that the Law, as ye sow so likewise shall ye reap, is the basis of all social distinctions. Each individual comes into physical expression and into a particular environment which is just as much a part of the individuality as is his physical body, and even more so; and in the consideration and interpretation of a natal chart, it is an impossibility to separate the native from his surrounding conditions. They constitute a part of himself and are those physical forces brought together for his requirements. Therefore the environment into which he or she is in must be taken as a part of themselves and all must be summed up in one grand total.

Then we arrive at an important point, as it follows that any change in environment will likewise denote a change in the native. This does not refer to every change of residence, traveling, etc., as these may not necessarily change the environments to any great extent; nor do they change the individuality or inner nature, but to great and

permanent changes in the personality of the soul's manifestation, as from a condition of the common people to
one of prominence, and the opportunities for the exercise
of power and authority. These will be the natural result
and follow as in harmony with the law, and will be in
exact manifestation with the growth and unfoldment of
the soul itself, and it must have utilized to the fullest
extent these opportunities that are afforded by such environments ere surroundings permitting greater power
and opportunities will present themselves.

The objective thought may be exercised constantly, but
will be of little consequence in the natural progress of the
individual until the soul from within has grown beyond
and superior to its environments. The great majority of
mankind are naturally seeking to find the way by which
they may best overcome the obstacles by which their environments have handicapped them, all for a purpose, too.
There may be a way in which this knowledge can be
applied in the every day life, and it is certainly a truth
that knowledge that will not admit of practical application
is really valueless, consisting of mere husks and shells,
upon which the soul in its effort to attain consciousness
and wisdom, may feed in vain, and this is in fact the condition of a great portion of the human race to-day.

All this must be changed. New thoughts must be
engendered; new aspirations must be pointed out and the
manner by which these may be attained set forth, for
surely there is hardly an individual but desires to find
success, and to get along in the world and create for himself an environment by which greater opportunities could
be given him for the exercise of power and also a greater
length of time to work along those lines most desired and
in harmony with the soul's desire for knowledge and
experience, and thus cut off the weary monotonous drudg-
ery, as it presents itself to the minds of the majority.

Then the question arises, What can be done? What
shall the individual do in order to attain this realization,

this new environment and power that will give opportunities?

This can be done providing he or she wills and acts in accordance with the will. First of all, dismiss all feelings of discontent; all thought spent in vain regrets. Lamentation at the hardness of our lot is a loss of power.

The astral form receives and preserves the manifestations of things visible and invisible to the mortal sight and transmits by a series of apparatus naturally adapted for the purpose, the atmospheric impressions and various influences which at a given moment are in the whole planetary system interblended.

The word influence and flux are derived from the same root word, *influo*, that is, to flow in upon. Thus planetary influence refers to the flowing forth from the planets of ethereal influences. The planets are composed of the same elementary parts as the Earth, and it is by a natural chemical action that this ethereal flow occurs. In disease the very name *"influenza"* implies some influence on mankind through the atmosphere, and so outside the surface of the Earth.

There is nothing indifferent in natural law. One stone more or less on the way may cause failure or success or alter the fortunes of the greatest men or the greatest empire, and much more then the position of a particular star cannot be indifferent to the destiny of the individual who is being born and who enters, by the fact of his birth, into the universal harmony of the astral forces, and thus finds his natural environments already prepared for his unfoldment and progress.

There is a loss of energy and power for those who stop to lament and magnify their troubles, wasting power that is essential to their success. Then this must be eradicated entirely and we must learn to take life as it comes in a joyful spirit, realizing the great purpose back of it all, and not looking upon it as an obstacle to advancement and unfoldment but as an absolutely essential manifestation of the astral forces for our special benefit, and that this is just exactly what it is, there is no question.

There must be certain knowledge gained from the environments in which one finds himself and the lesson to be attained must be learned by heart, so to speak, ere there will come a change of conditions.

Humanity the world over, at least a large majority, love changes and varieties in life, and many are placed in a position where they are compelled to tread the weary round of dull routine day after day, week after week, aye, year after year, and naturally will at times feel disposed to resent it, and wonder why they should be compelled to perform certain labors which are distasteful and disagreeable to them. The answer to this query is simply because the soul needs strengthening in a certain direction or quality which is latent, and the two qualities most essential to the individual's progress are patience and perseverance.

It is essential to put the whole energy to work and to act with a will until the lesson is gained, no matter how annoying or distasteful, and if the individual would only stop and realize that in all manifestations of the forces about him there is a lesson to learn, and consider his duty and obligations, not as a debt to be paid with bad grace, but as a means of purchasing success, not wasting energy in building castles or making plans that lie so far beyond, they may never be carried out through lack of opportunity, and that very lack of opportunity may arise through the lack of power to eradicate the present environment, brought on by wasted or misdirected energy.

The individual must learn to concentrate all his forces into the present moment, pouring them into his present duties, for he is living in the eternal now, and as the soul grows too large to be contained within the limits of present environments, this environment will become disintegrated and as his stars progress, fresh opportunities will be presented, and then, if the power has been well directed, it will be ready to go forward and accept the opportunity; but the opportunity will not be presented until the soul is ready, and it will not be ready until it has learned its lessons from present environments, and it will not learn its lessons unless the energies are directed to the present, with its many tasks and opportunities.

There is much darkness, ignorance and misconception to be overcome; the light of real truth that shines down into the soul through the power of intuition, belongs to a higher and immortal existence and must not be considered as belonging to the physical form, and the reflection of that great esoteric light is best known as intellect. This endows mankind with thinking, reasoning faculties from which proceed the personality.

As all things are given into man's possession to make use of, either for good or evil, so likewise the intellect may lead him into the paths of perfection and advancement, or if the environments are otherwise, the effects will be apt to manifest in just the opposite manner

So long as the individual holds the divine light before him he will have a guide which will surely lead him from the darkest clouds of ignorance upward to the heights of wisdom. It is the fountain of all absolute knowledge out of which emanates all wisdom, but if the individual detaches himself from this source of truth and knowledge for good, and employs only the lower side of his mind, the understanding will then become dazzled by the self-conceit which is apt to decoy him into the quicksands of illusion and deception, and, at the same time, the emotional fire degenerates into the flames of an inferno.

The fundamental or principal cause of all personal knowledge is curiosity, and without this curiosity and desire to know, there would be no attempt made to observe or cherish any great longings in one's thoughts; as, for instance, had the mythical Eve not been curious to taste the forbidden fruit of the tree of knowledge of good and evil, the first man would not have fallen from his condition in paradise and thus stepped his foot upon the lowest round of the ladder of individual progress and unfoldment, and thus from curiosity arises a desire to satisfy it.

This is in harmony with Natures' law for the reason that the motive is not only good but intellectual, but, as in the case of the mythical Adam and Eve, a source of trouble may arise in attempting to penetrate the arcana of Nature and the mysteries of the universe, although the proper pursuit of scientific knowledge cannot be otherwise than uplifting and elevating.

The soul is filled with high aspirations and compels the admiration of that divine creative force which first called the universe into objective existence; but on the other hand, if the spiritual view of the soul is obstructed, in the place of reverence for being, there remains then only the throne of matter, and if the divine law of wisdom is ignored entirely, and blind ambition and applause recognized and sought as rulers, then the evil of conceit holds sway, and from this emanates a class of individuals who may have sought truth but who, deserted by good, imagine themselves to be perfect, and where all sense of the spiritual, sacred, and sublime are lost, together with every religious instinct, they are then no longer able to distinguish anything spiritual or elevating, either in the Nature law or in their fellowman.

The great desire for the satisfaction of their scientific curiosity and personal ambition guides them to Nature's defamation as well as their own degradation and humiliation, and even in physical discipline, science could make more rapid progress, if mankind would stop to grasp and comprehend the physical aspects of Natural law. That force which lies back of every mechanical working power is intelligence, the very source of that power, and if the spark of divinity be thrust to one side, in which is embodied perception and self-consciousness of life, the resulting ignorance is the cause of needless griefs and injuries that are not easily eradicated.

It is not possible to ignore that divine spark, that is manifesting in some degree of intensity, without injurious results to the welfare of the individual and naturally retarding the progress for a time. There is the greatest injury done to science by physiologists, biologists, anatomists and others who are buried in dogmas and creeds which stand in the way of truth and science, and ever will, until they are awakened from their condition of lethargy. We refer more particularly to those individuals who call themselves scientific investigators who take it upon themselves to meddle with the great manifestations of the Law, the phenomena of life divine, and who have really lost all perception of their import and spirituality; as, for instance, the physician, in his studies of man's organization, devotes his entire energy upon the material body and its functions, and for the reason that he does not realize the fact that the material body is but a vehicle, a sheath for the spirit, which is in turn a sheath for the soul, the temple of the divine man; he will naturally proceed without regard or respect for the individual himself and for the very reason that he is only interested in the objective, the concrete and what his material vision can behold, also is possibly unbelieving as to the existence of a soul, he merely contemplates man's material body in which the life forces have manifested, as a mere subject for investigation and examination in the dissecting room. He no doubt feels that he is to be a benefactor to his fellowman, and derives some degree of satisfaction in such investigation, and he does not hesitate to pursue and trespass upon the living form whose life functions have not been extinguished.

It is then but a step from the observation and explanation of Nature to the horrors of vivisection, for if this love of cruelty is once aroused in the name of scientific investigation and all sense of compassion lost, a person so degraded and manifesting in the lower animal instincts can scarcely be prevented from perpetrating and committing crimes to satisfy his desire and craving for knowledge when he stands in no fear of his country's laws and in fact is upheld and protected by them

There is no crime, no matter how horrible or unjust it may be, for which some excuse may not be set forth in consequence, and for this reason the cruelty of vivisection is not only publicly flaunted, but public support is solicited in its behalf under the representation that from such cruelty and terrible practices will some day result discoveries of something that may be of interest and benefit to others.

Such experiments are useless, as can be demonstrated, though it seems a most difficult task to convince those who have no perception and consciousness of the sanctity of being and the divinity of man, that there is no question that mankind is not possessed of the right to sacrifice the higher life principles to the lower manifestations of matter. Such conditions must be changed ere the world can hope to eradicate crime and the taking of human life.

Inmates of poorhouses, hospitals, asylums and prisons are subjected to these scientific experiments, their material bodies infected with the germs of horrible diseases, the results of which are most cruel sufferings. These are but the fruits of a mad haste that oversteps the limits of the natural law. The only prevention is to illuminate science with the light of this divine science through the recognition of every higher aspect of humanity it teaches.

CHAPTER II

There Are Seven Principles Associated With the One Central Spirit of the Solar System.

As we have learned, there are seven principles associated with the one central spirit of the solar system and from the most exalted manifestation of the consciousness of the Logos down to the most minute atom, the influence or effect of these seven principles is in active expression.

In a study of so vast a subject as the solar system it is quite difficult to tabulate the various qualities as governed by each of the great lords or spirits, and the only course to adopt for the purpose of clear expression, is to deal directly with the influence of the planets and the two luminaries (the Sun and the Moon), and then seek to realize and to comprehend their great and mighty influence on the higher planes of life beyond the physical.

In considering these seven principles we find only five in active manifestation, two are still concealed in infinity itself, as the universe is now in the process of evolution, and thus far has only attained to the sixth stage or round in the ladder of progress; they are the ruler of air, the ruler of water, the ruler of fire, the ruler of Earth and the ruler of the ether vibrations.

It will be well first to make clear to the understanding regarding this influence of the physical planets which bear the names we have applied to them and by which they are generally known. The physical planets themselves do not affect humanity direct or first hand, but the spirit or principle back of that planet is the source or cause of the influence that affects all matter existing in the universe.

We can best consider the physical planets or globes as the dense bodies or vehicles, similar to the physical body of man, through which the soul and spirit is finding expression. Thus the physical globe itself is the vehicle of expression of the influence back of it, which principle is in turn under the influence of still higher and greater powers.

These influences are in reality vibrations set in motion by the principles, aye, intelligences back of the physical globe itself, and from the highest to the lowest this vibration is in constant manifestation, all forms and substances being subject to these higher vibratory laws. It may be more clear to the mind if we speak of these intelligences back of the physical globe or refer to them as individualities,* who one day obtained a physical experience upon that globe or planet with which they are associated, and all the vibrations connected with this planet and emanating from its sphere of influence, are included in this consideration, and so far as the real effects of these vibrations upon other planets or globes is concerned, the planets act as the focus or distributing agents.

The vibrations of each planetary sphere of influence both in the highest as well as lowest aspects, possess many subdivisions, their vibrations affecting not only all so-called solids, but liquid and gaseous elements also extend their influence to different states of ether beyond those that the mortal mind has any understanding or comprehension of, as, for instance, the intelligences back of the planet Saturn have rule or influence over the mineral kingdom, the objective or concrete condition of matter, and if we would think of the Saturn vibrations in association with the liquid state we should consider ice as coming under this influence, also the north and east wind, so far as the airy condition of matter is concerned. It is considered that Saturn is malefic and this is owing

_____

*See Vol. IV., Pages 109 to 118 inclusive.

to the fact that this planet brings all things to the crystalline condition, and from this standpoint his vibration is considered evil, and it is well known that Saturn precipitates and binds all things over which he has influence.

In further considering these intelligences who have influence over the various globes and planets of the universe, we must comprehend that we are dealing with individuals who are on the way to perfection and associated with each separate planet are intelligences who have developed various qualities and virtues that are essential to the attainment of perfection; but in each instance we shall find that the intelligences back of the planet itself, are manifesting in those conditions and qualities that are attributed to the planets themselves. To be sure, we on the Earth planet consider the influence of the planets only so far as their effects upon the Earth are manifested; in other words, we judge of them from our limited knowledge and education, and not from a universal knowledge of their nature, and it is well that we confine ourselves to this limitation, for it is never wise to go beyond the comprehension.

Each of These Intelligences Manifest in Harmony as One Great Whole, but Are Distinct Individualities.

As referred to, Saturn is a planet of limitations, at least as far as his effects upon this Earth are concerned, and his vibrations have rule over all solids, from the finest crystal to the hardest rock or mineral substance. It is well known that Saturn has control over the human framework or the bony structure of all living creatures.

It will be more easily understood if we consider each of the seven ruling spirits to be a modification of the consciousness of the solar Logos, and that each of these intelligences, of which there are many, who manifest in

harmony as one great whole, but still distinct individualities, have control over a certain plane in the material,
also in color and sound as well as those principles and
qualities that are co-existent with the subjective and ideal
planes of life.

We can now simplify this if we consider three of these
great forces of the five in nature that are in manifestation,
and using the ordinary terms of the planets to designate
them as we have learned them heretofore, that is, Saturn,
Jupiter and Mars; but referring to the intelligence back
of them as the cause of the influence as manifested by
each.

We have already considered the influence as attributed to Saturn, as it governs and has control over the
whole of the mineral kingdom in its most concrete condition, and most important, too, the framework of the body.
The term or word used by the Hindus to express this,
namely, Tamas, makes it quite clear to the mind, and
correctly interpreted really conveys the true idea of this
planet. In its English translation the interpretation as
inertia, stability and resistance, it will be found that
this planet, Saturn, gives the solidity required by all
things in order to make them enduring, and thus, in this
consideration, Saturn has rule over the lowest and most
concrete of all atoms or substances, and then again, on the
other hand, it is concerned and associated with the very
highest manifestation of the one great life.

Then again, for the purpose of obtaining a clearer comprehension, concerning the vibrations and influences of
the planets, we may consider the dual evolution of life
and form, that is, spirit matter, though in reality, as we
have learned heretofore, the two are inseparable at least,
to the present condition of understanding.

Each separate and distinct planetary vibration may be
considered to consist of three distinct aspects of itself, and
these three states or aspects are known to the Hindus by
the terms Tamas, Rajas and Sattra, and translated,

signify, Slavery, Service and Mastery; and at certain stages of the life evolvment only one of these three aspects will be in active manifestation.

We may now begin to realize that what is generally considered as the active manifestation of Nature's law is the result of intelligences who have unfolded to that height of wisdom that they are particularly adapted to guide and control those influences or vibrations over which they have power to send out into space to perform their mission. It is the intelligence back of all substance that causes the proper manifestation of that substance as a vehicle of expression for certain forces most essential to the evolution of the whole.

We have learned that in the mineral kingdom the Saturn vibrations are most pronounced, and during the great cycles the intelligence back of this planet will find expression through the most dense substance to the finest crystal and finally manifesting in the lowest lichen. In dealing with metals we can see a specialization of the Saturnine rays manifesting through lead, a metal that is known to be especially ductile and enduring, and that which this planet has rule over in the substance or objective side of life, the vibration or influence manifesting through that substance will retain in it inertia, resistance and durability as the ultimate goal, for that vibration is found in permanence, and if we but study closely this particular vibration on the life side of evolution the intellectual manifestation of it, we shall find that the consciousness in its association with the objective, will always be seeking these qualities.

Humanity generally picks up the qualities of rigidity and stability which have been acquired through past ages while the elemental essence was on its way downwards into evolution in the objective forms which it takes possession of when drawn into physical manifestation. It is a fact that there is an evolution of the life, as well as the form, going on, and thus making the vehicle of expression more clear and distinct. It is true that the consciousness of every human being does not exist in the first attempts of the ego to gain experience upon the physical globe. It may sleep in the rocks, dream in the animal and finally awaken in the human.

The Mystical Side of the Planet Saturn.

We may learn much of the mystical side of the planet Saturn from its name, as name is power and expresses the quality of life. We may divide the word into two: *Sat* is from the Sanskrit and interpreted signifies Be-ness, and it is that divine essence which is. It does not exist for it can only be made manifest, so to speak, through the *urn*. The symbolism, therefore, of this planet, signifies spirit substance in its true sense, as will be observed by the cross over the half circle or crescent, representative of the Moon.   .

In the beginning of Saturn's influence or vibration, the qualities inertia, resistance and stability are more powerfully manifested than at any other time, they will be found embedded in matter where all is naturally darkness to the spiritual side of the life, and inertia brings out the quality of resistance and stability, thus the life of spiritual finds the power to resist and endure, then will finally learn through these very limitations the necessity for the expression of those virtues of peace and economy.

Satwa is the Sanskrit word, and interpreted means goodness, purity and understanding, and though great periods of time will elapse, nevertheless the qualities of slavery must finally give way to mastery. In the animal and human, Saturn governs the bones, the spleen, also melancholy and sorrowful emotions and the individual mind in its most concrete and objective expression as well as the meditative, reasoning and contemplative condition of the individual; we may thus find much food for thought in the study of the various states of consciousness, as vibrations of Saturn.

Within the limitations of time the heptad or number three is important for the reason that it is the key number of the manifested universe. The Logos, as we have seen heretofore, manifests as a duality, a two in one, but when the egg of the universe, so to speak, is formed and differentiated, its results are ever manifested seven-fold whether stated in terms of matter or consciousness, seven esoterically, seven exoterically.

The first great modes of consciousness following immediately after the divine consciousness are seven and they have been spoken of as the seven Gods or seven Logi. They are subordinate to the One, standing to him as the seven rays of the spectrum stand to white light or as the seven notes in music compare to sound, or the seven planets in their influence harmonize with the Sun and Moon. Number seven is really a seven in one and similar to the number five. It is derived from the monad for the reason it cannot be formed by multiplication. It is represented by the seven pointed star and the interlaced triangles with the centralizing point. It may be divided into different forms; the most important, however, are the triad and tetrad.

In the ancient writings the seven planets are often referred to, though it was the seven Gods and the principles for which they stood that were meant and not the physical planets themselves, and no matter how many planets there may be on the physical plane, they are all grouped into a seven-fold classification.

There are seven schemes of evolution going on, each being concerned with one, and some with more than one physical planet, and with several as you would term, invisible globes. These seven Gods are the highest and most spiritual administrators of the commands given to them by the Universal One, the Logos, and as we may term, Creative Deity, though each one taken alone is an actual existing intelligence, and is also the head of a hierarchial host of intelligences of all grades, ranging from the highest archangel to the lowest nature spirit.

We are seeking to use terms as simple and comprehensive as possible, and there is no form of life or consciousness anywhere in our solar system that is not related, and belongs to some one of these seven groups, and is in some mode, high or low, a manifestation of the life and consciousness in harmony or affinity with the seven Gods or guiding spirits.

The Hindus speak and write of these seven Gods as Suras, sometimes as Devas. The Hebrew, the Mussulman and the Christian speak of them as angels and archangels, thus making a distinction between the higher and the lower intelligences and their planes of development. Zoroaster wrote of them as the Feristhas. The Chaldeans considered the seven Great Gods or powers that were in control of the seven physical planets; five of these were in active manifestation so far as the mortal consciousness was concerned, while the other two were not realized as yet by those in the physical expression, as, in the process of evolution, only five stages or planes have been traversed. The Hindus name these five as we have heretofore described.

These seven Gods take up the work of the one Creative Deity and establish seven separate centers within the universe, and each of these is again subdivided on a septenary scale, and thus from the one Logos, the Creative Deity, we proceed to the two, from the two to the five, from the five to the seven, and again from the seven to the forty-nine, and so on, and on to the infinitely subdivided beyond mortal comprehension.

These seven intelligences or Gods are the rulers and controllers of the seven elements, five of which we have heretofore considered; also the seven conditions or methods of motion in matter; also the seven planes, the seven schemes of evolution as well as the other septenary groups in nature. Each planet really contains and possesses the powers of all the others within it, though naturally its own special character will predominate, and the same condition is true of each of the twelve signs of the Zodiac, and with all substance the whole is ever reflected in its parts.

In the Zodiacal sign Libra, we have corresponding reflection of the third or the seven cosmic planes, the third point of the manifested triangle, and what may be termed the Nirvanic plane, and also the third reflection of the Atmu, and in a lower expression, Manas, the human ego. On this plane is the abode of the Kimar, as this refers to those individuals who have attained to Omniscience regarding all that belongs to the realm of Maya and are naturally under its sway.

The Kimaras are a divine hierarchy of highly developed individuals who played an important part in evolving, in our infant humanity several millions of years in the past, the germ of the mind, self-conscious intelligence, the human ego itself, and this condition is represented by the sign Capricorn.

Saturn, the significator of the individualized and self-limited ego, finds his natural home in the sign Capricorn, the sign representative of the ego, and Saturn finds his exaltation in the sign of Libra, the balance and the plane reflected from the Kimaras. Manas, or the human self-conscious mind, was assisted and fostered by these intelligences.

The human race having attained to the point of receptivity, the Kimaras reflected out to mankind the spark of the higher consciousness, the balance needed for the evolvement to that higher state of being. It is true that these Kimaras from other globes, who had completed their own intellectual evolution on those globes, came to this Earth planet for the purpose of aiding in the evolution of the egos attracted to this globe, and really represent the spiritual fathers and mothers of our humanity.

Libra is the home of Venus, its night house, and thus we have the relation which exists in its affinity with the Earth, also the same relation exists as regards Venus, Saturn, Libra and Capricorn, the Zodiacal counterpart of the occult expression. In the Orphic theogony Saturn belongs to the neodic or intellectual order, the third below, immediately following in order to which the planet Urania belongs. With Saturn is associated Rhea, his consort, also Jupiter and Oceanus, the separative deity, the latter referring to the next sign above, that is, Scorpio; but conjoined with Rhea there is another triad termed the Kopos, meaning virgin, for their powers are pure and virginlike. There is a similarity here between these and the Kimaras of Hindu mythology. The word Kumdra also signifies virgin, and who were also seven in number. The characteristics attributed to Cronus Saturn in classical mythologies can be traced in numerous myths belonging to India, Greece and Egypt in various manners of symbolism, and may finally be tested by the student in planetary and Zodiacal influences to-day.

It would take volumes to properly trace out all these lines of connection, but we will consider a few important thoughts on this subject.

Saturn and Libra Stand for an Individualized Center of Consciousness.

The planet Saturn and the sign Libra stand for a definitely individualized center of consciousness, and considered cosmically only one such center is possible, that is, the divine center of infinite self-consciousness which orders, governs, controls and animates the whole universe and imparts to it its unity, for without unity, there can be no universe, and Libra stands for this unity or its expression in its intellectually self-conscious mode.

In considering humanity from a microcosmical interpretation, each and every human being is comparatively a small universe within the aura or atmosphere which he finds expression, and the manifestations of the consciousness within, will have influence and control over certain atoms, forces, etc., belonging in that individualized system. Now the planet Saturn in the sign Capricorn signifies an analogous state or condition in the individual to that for which the sign Libra compares in the universe, that is, the limitation of intellectual self-consciousness. It was proclaimed by Proclus and Helmus before him, that the planet Saturn alone deprives heaven of the kingdom cutting, and being cut off, and in this we have an idea of the power that first creates separate units, that is, human intellect, and then finally unifies them, that is, divine intellect, and this is just exactly what Saturn does, in the so-called myth. He devours his own offspring, which signifies that having first created them as separate sparks in their own lower expression of the concrete, he finally evolves them through his natural qualities to that higher and finer degree of his own consciousness, and thus the individual attains that divine condition, and is evolved through the evolutionary process in harmony with other great divine minds. Though Jupiter escapes being thus devoured for the reason that he stands for an intermediate state between the two which is termed Buddhic in the individual, and is concerned with the qualities as signified by the sign Saggitarius in the Zodiac, that is a state or condition of duality in unity where two separate individuals may harmonize together as one without either losing their individual consciousness, which may seem impossible and apparently contradictory, but is actual and possible in nature, however incomprehensible to the mortal mind.

Socrates terms Saturn insolent for the reason that he

deprives his father of the kingdom, and in reality we find in Urania the connective power in the same way that Saturn is the separating power. We find a correspondence of the Sanskrit Rapasa, that is, the active moving and energising Guna or qualities, and we also find that the sign Libra corresponds to the Tattras which refer to the various modes of motion in matter, and brought about by varied modifications of the divine intelligences who are therefore most active agents in differentiating the cosmos, and finally as bearing upon this same separating quality.

It will be observed that the line of the horizon in the horoscopal figure is completed or finished in forming the cusp of the seventh house, which is Libra, in the radical or natural position, and this is the diameter of the circle which divides the lower from the upper and thus differentiates the circle.

The circle alone stands for zero or naught, but the point within that, marks the center of the circle or the diameter. This line that forms the ascendant and descendant represents separation and limitation, and the two taken together form the number ten, which represents and signifies the whole universe.

Then the second line or diameter, that is the line that forms the cusps of the fourth and the tenth houses and runs from the Nadir to the mid-heaven, starting from the lowest point in the figure, it has its ending in that point in the circle which corresponds with the sign Capricorn, the tenth sign of the Zodiac, the home of Saturn, the planet of separation and limitation, and Saturn finds his exaltation in the sign Libra, the tenth from the tenth.

We desire that all we may say will find root in the minds of humanity, and open the way to a higher, clearer understanding of the creative forces in the nature law, and their modes of operation in the evolutionary process of the universe; and that from these writings mankind may receive a correct understanding of past experiences of those who have heretofore traversed the beaten path from the physical to the higher spheres, and from these truths we may better realize the significance of the ancient myths and allegories that have been written of so extensively, but at all times material ideas of man have predominated.

The myths of Promethus and Satan and of the fall of the angels and also of man. The Nirvanic planes may also be considered here, that is, the three signs of Leo, Virgo and Libra taken together and considered as one so far as their influence upon man is concerned, and may be termed Nirvanic. Nirvana is the heart, so to speak, of the universe, from whence all its life currents emanate. Thus the great *Influo* or vibration comes forth the life of all, and there also is it indrawn when the universe has reached its term. There is unveiled the supreme goal to which all humanity are evolving.

All the characteristics that are known to be associated with the sign Libra and the seventh house, correspond with all that has gone before, for unity is the keynote of the whole.

Naturally in the physical expression we observe the objective manifestation expressed in love and marriage, a unity with the object of desire, and should be love, but is often passion instead, though in itself, this unity is that of all beings with one another and also their creator, so to speak. It is in truth, existence raised to a reality and exactness incomprehensible to those who realize alone through the life senses and the mind in the objective.

As the lamplight is to the brightness of the noonday Sun, so is the Earth-environed consciousness to the Nirvanic planes.

In the first thought this condition might seem to conflict with the limiting or the separating tendency associated with the Saturn vibration, or the neoric order of things. It does not, in reality, for the neoric or Saturn limitation is cosmic and not individual. It is in reality that limitation and separation that marks off one universe from another, and it is in truth to the whole universe what individual consciousness is to man, and if we imagine an individualized soul, no matter in what plane, the manifestation is taking place, whether in the lower or the higher planes of consciousness.

This very fact of individuality naturally implies two apparently opposing forces or qualities, that is, separation and unity. The separation marks off a definite center of consciousness, while the unity synthesizes all things within the boundary of that center, and then in the Nirvanic consciousness the center has been expanded, as it were, until its limits harmonize and compare with the limits of the universe itself.

Those individuals who have this Libra influence strongly manifested in their natal chart of birth, will often show in their lives and characteristics this apparent contradiction of unity and separateness, as they are independent, erratic, changeable and impulsive, and at the same time they possess a strong inclination to associate themselves with other individuals, often with some one person, for work as well as pleasure, and they find it difficult to live alone or work alone. In fact, they cannot do so and be really happy, and they are most contented when married or engaged in partnerships. The interplaced triangles with the central point form a harmonious symbol for the union found in Libra and the seventh house.

This subject requires careful study and consideration to comprehend, as we are dealing with forces, vibrations and intelligences that are far beyond mortal understanding, though some ideas can be gained as to the conditions that exist in the spheres and planes of life beyond our own. Once this is understood, the human race will more quickly evolve to that higher consciousness. They will cease their laborious task of delving in the concrete and lift up their eyes to the life that surrounds them, which is tinged with the lessons that will prepare them for the next step of evolutionary progress. The sorrows and afflictions of life will be banished in the eternal joys of developed consciousness, a consciousness of being that is tangible and real, not a mere shadow and illusion.

As we have heretofore described, the symbol that stands for the sign Libra is well adapted to express those qualities and characteristics that naturally belong to that sign. Libra being the seventh sign, finds affinity with the seventh sign esoterically, that is, Pisces, in which sign, Venus, the ruler of Libra, finds her exaltation.

If the planets be arranged around a seven-pointed star in their order and proportion of the atomic weight of the metals with which they are alchemically associated, every alternate planet around the star will fall in the order of days of the week, and the vibrations of the star pass from one planet to another in the order of the motion of the the planets in the Zodiac.

It will be observed that seven as a number is made up of four and three or five and two or six and one; and if for these numbers the corresponding planets be substituted we shall find that such is the case. In the Hindu system which we have heretofore considered to some extent it has more or less direct influence over the affairs of this seventh number, referring to partnerships, love and marriage. Thus in the Mercury, signified by the sixth sub-period of Mars signified by the first period, the significations are that the native is liable to marry at this time or enter partnerships, and the influence argues well. And again, for instance, in the Mercury signified by third sub-period of the Moon fourth period, the native will be fortunate and find amusement through affairs of love and children.

*Resemblance of the Symbol that Stands for Saturn and
the Figure Seven to the Sickle or Scythe Used
by Father Time.*

It will be observed the resemblance of the scythe or
sickle as used by Father Time to the figure seven, also
to the symbol that stands for the figure Saturn, and beginning with the Earth, Saturn is the seventh planet and
finds his exaltation in the seventh house Libra. The
planets visible to the mortal vision, at least in this solar
system, with their invisible globes, are gathered up into
seven evolutionary schemes or chains, and these find their
correspondence to the seven signs within the egg signified
by the sign Cancer.

We may further consider this arrangement by comparing the planetary chain with the Zodiacal signs in
their affinity to each other: First, the Sun with the sign
Leo; second chain, the Earth, which also includes Mars
and Mercury, and corresponds to the triad signs, Virgo,
Libra and Scorpio. These were considered as one by the
ancients. Then, third chain, Venus and her day sign,
Taurus; then the fourth chain, Jupiter, and signified by
his sign Sagittarius. The fifth chain, Saturn, signified by
his sign Capricorn; the sixth chain, Urania, signified by
the sign Aquarius; the seventh and last chain, Neptune,
Acasia, and one other planet yet to be discovered, and
these last three find their affinity in the Zodiacal sign
Pisces.

We are now dealing with the higher vibrations of this
solar system. The natural destiny to be realized by the
human race in their evolvement from the lower planes of
life into those higher planes, where the consciousness
really begins to comprehend what is beyond it, and as it
comes into possession of one key after another, it may in
turn unlock the doors that stand between knowledge and
wisdom; the mystery of life will be unfolded; Isis will be

unveiled, and the heavenly light will penetrate all-where. This will have an effect upon the lower planes more especially.

All things must move in perfect harmony in the law, as, for instance, if the vibrations beyond in the higher planes should be suddenly merged into the atmosphere of the lower spheres, chaos would be the result. No, the law must be manifested to all according to their plane of understanding, and not beyond a certain rate of vibration could they be taken without producing some abnormal effect upon the individual; in fact, this is the case in many charts of birth.

In considering the planetary chain, the asteroids were not numbered among the planets, however they come under the second planetary chain, and are really under the influence of Scorpio, as their influence corresponds to this sign at the present time, that is, taken collectively, though in fact they are what might be termed unused matter in a sense, and are awaiting further development, and it will be observed that the sign Scorpio, under which influence they are now manifesting, is the sign that provides material for human development, and similar to this sign, the evil influence is disintegrating and destructive, though the good influence is energizing and life-giving, and as they evolve, these qualities will be manifested for good along the lower planes of life and they will act as the vehicles for transmission of higher vibrations from the planes of life above to those planes below.

The seven spheres of evolution are associated with the physical planets themselves, and more especially with their invisible globes that belong to them in a similar manner as we have heretofore described, the invisible globes* associated with this, our Earth planet; but still beyond this scheme of seven, there are still seven other evo-

---

*See Vol. IV., Chapter 16.

lutionary schemes going on upon these invisible globes
that are not directly associated with the physical plane of
life, though all are related and associated indirectly as to
one grand whole.

These are, then, really concerned with the higher
orders of evolution, and with the ultimate perfection of
the life of the solar system as a whole, when all the septenary schemes have completed their cycles and they are
in comparison to the seven invisible planets above the
physical planets themselves.

Thus it will be seen that in the evolutionary scheme,
seven is the complete number on the physical and spiritual planes.   Thus we have in the signs Aries or Libra,
inclusive, the evolving chain, each one in their order,
manifesting their own particular nature and qualities
according to the desired needs.

According to the arrangement of the planetary chain,
our Earth planet is under the sign Virgo.   This is significant when the nature and qualities of this sign are considered.

In the ancient myths of Mother Earth, Terra and Car-
los, also Gaia and Urania, the all producing, the mother,
fall into their proper places in the sign Virgo.   The term
Earth, refers not only to primordial mother substance, but
in a lower concrete expression to our globe, and would
then be associated with the sign Virgo, which is numbered
as two in the seven.

CHAPTER III

The Signification of Numbers and Their Relation to Astrology Demonstrated.

It is a truth that the fate and so-called fortune, as well as the character and environments of any individual could be expressed in terms of numbers, and the more highly developed and specialized the individual, the higher and more complex will be the number necessary to express it. In both instances, the cycle returns upon itself and comes into higher perfected unity.

The simpler numbers are more comprehensive to the mortal mind. The procession of numbers from unity to multitude, corresponds to the differentiation of the nebulæ of a solar system into Sun, Moon and planets, also to that evolution by which the spiritual monad of a group of animals becomes more and more separated and limited within its own environment, until individuality is attained, and thus individuals stand to the all embracing spiritual monad as the higher and more complex numbers stand to unity itself.

In considering the evolution of the soul, it will be observed that it gains greater and more glorious powers until it ultimately attains to the condition of at-one-ment with the Deity, or the source from which it started out on its long pilgrimage, and in the procession of numbers beginning at unity, each number becomes more complex and specialized until the whole returns to the higher unity.

All human life and all mundane affairs proceed in cycles, and this is demonstrated by a careful examination of the world's evolution. Many of those cycles are quite simple in their manifestations, while others are most complex, even beyond the conception of the mortal mind today, as there are cycles dealing with the Zodiac and also of the planets, of which we have referred to in other writings, and still beyond these, there are the cosmic cycles of which it is difficult for the mortal mind manifesting in the five lower planes of life to realize or comprehend. In the consideration of these cycles we shall find that their influence or manifestations are varied according to the nature of the numbers involved, as numbers change their apparent characteristics with the point of view of the observer, and upon the physical plane of life they are generally used at random to signify units or groups of units, but when considered from the higher planes they really become the manifesting expressors of the working of Deity in the universe. The cipher or naught represents in its circle formation the inexhaustible source of all numbers unknown to the mortal mind and incomprehensible to them.

It is not in reality a vacuum, an empty nothingness, but the permanent unchangeable basis of the manifestation of changeable things. It never exists but always is, no matter whether there be a universe or whether there is none when the cosmos comes to that volvement that plane after plane is rolled up, as it were like a scroll, when all beings evolve to that infinite condition incomprehensible to mortal mind, and enter that absolute light as one harmonious whole. It is the naught, the eternal one being, who have their eternal work to perform in the formation of worlds, planets and systems and the evolvement of the egos who go out into space from unity to find expression in the lower concrete forms of life, finally attaining to consciousness in the human expression of man and woman, and thus the great evolutionary process goes on from cycle to cycle and the condition attained when all, as one, evolve into harmony. Then it is the symbol naught that aptly expresses the condition of mankind.

There can be no great distinction drawn that will be very clear to mortal mind, but we may consider the monad to represent a tendency to unite into one centre, and as it were, become definite out of the indefinite universal all. In truth no real line of distinction or definition can be made, for the monad merges into the naught and really becomes one with it in the dissolution of things and finally re-emerges at the dawn of a new universe to create and formulate in harmony with the eternal evolutionary process.

There is in reality neither first or last, for all is one number manifesting from the nought, and as applied to a differentiated series of many, the monad or number one is merely the first number and the typical odd number, though when considered in an esoteric sense, it is neither odd nor even, and may hardly be termed even a number at all. It really signifies the first expression in the process that brings about the manifestation of all the numbers both odd and even. Esoterically it is the divine unit, not one in a series of many, but the absolute monad that comprises and embraces all, as if there were no universe. It ceases to exist but still is, that is, merged into naught.

If there is a universe, this figure one contains within itself potentially all the numbers of differentiation which go to make up the universe, though it still remains a unit. It contains all knowledge, but to the mortal mind is incomprehensible. Knowledge implies duality, that is, the knower and the known, while in the monad we have unity. It is neither light or darkness, neither life or death, neither good nor evil, and it is neither spirit nor matter, for in absolute unity no duality of any kind can exist. It will lift up into the plane above all others, and is superior to them all but it is the potential source of them all.

As we have learned heretofore, all things in the universe move in cycles, even motion starts from a point, passes along its arc, and then returns as a circle to the point from which it was created. In reality its progress is spiral though one round of a spiral may be represented as a circle.

The Number One is the Inexhaustible Fount of all Differentiated Things.

Thus we observe there are two methods of considering any number or group of numbers, that is, according as they operate on the downward and outward arc, or on that which is inward and upward in its tendency, and when these two arcs are represented in terms of motion along a straight line, the one will necessarily be the reverse of the other, that is, the outgoing from A to B and the inflowing, the reverse, from B to A; and in just the same manner the characteristics of the upward arc completely reverse those of the downward arc, and yet, for the reason that the motion is not really in a straight line, but in a circle, although metaphysically in the same direction; always some common property must belong to it, all the time on its path of eternal progress. This is also true of all numbers. The number one always stands for unity, whether in the great infinite life or on the lower plane of the concrete form, though it varies in characteristics according to whether it is regarded as the source and origin of all things on the downward arc, or as the goal of all things on the upward arc.

Inasmuch as it is the inexhaustible fount of all differentiated things on the lower planes, whether speaking in terms of spirit or matter, or of consciousness, it must be regarded as creative or productive in its potentiality, though if we look upon it as the goal or end of all separated things, it becomes destructive apparently though not in reality.

In further consideration of the unity or one, it may be regarded as the universe as one grand whole, and in the science of astronomy it represents the nebulae from which, Sun, Moon, planets and stars are created, or it really signifies the Sun itself.

In the divine science of astrology it represents the Earth planet which unifies the Zodiac so far as we are concerned who are existing in the confines of this Earth planet, and it also represents the Zodiac when considered as one whole.

In the physical man it represents the heart, in the spiritual man it represents the divine consciousness, and in the earth it represents an internal central point midway between the two poles. It may be observed that just as the number one may be regarded as either creative or destructive according to the point of view, so also the first sign of the Zodiac, Aries, and the ascendant which are creative and life-giving, are linked with the Zodiac sign Scorpio and the eighth house, which stand for, and have rule over death of the physical, and thus are the positive and negative manifestations of the same planet Mars.

When separate units are merged into one, each is destroyed as such, although indestructibly preserved in the greater unity to re-emerge therefrom into a higher manifestation at the dawn of a new cycle. The outflowing current separates, and creates, and brings separate centres into being, while on the other hand the inflowing current unifies, and from the point of view of the separated self seems to destroy, though in reality the destroyer and the creator are one.

In the measure of time this unit, one, stands for any true unit manifesting in Nature's law, as distinguished from those units invented by mortal man for his own convenience, and this is certainly a most important distinction, and of these true unit measures there are three chief ones, the rotation of the Earth on its axis giving rise to the day unit is the first, while the motion of the

Moon in her synodical month is the second, and the revolution of the Earth planet round the Sun annually, signifying the year unit, is the third. These three units are by correspondence equivalent to each other.

It is possible to draw several conclusions of vast importance in directions from these measures of time. The symbol associated with this stage of cosmic evolution is the point within the circle, which is also that of the Sun, and it will be observed that the Sun is exalted in the fiery sign of Aries.

CHAPTER IV

By Self-Transformation One is Created Into Two.

By self transformation one is created into two. This number two is not a unit or monad, as is the case with number one; it is the dual or two in one dual in nature, for here separation has taken place, though like number one it involves two aspects, that is, one on the downward or creative arc, and that on the synthesizing or upward arc.

When on the downward progression from the one to the many, it signifies separation, while unity gives birth to duality as the first step in the process of evolution of a multitude of separated units, and in this consideration, this number may signify opposition and enmity, and this is the reason why it was considered by the ancients to be the number of evil influence; though on the upward arc it signifies the merging of two in one, and from this point of view signifies the union of pairs in their opposites for peace and war, for harmony and discord, for happiness and sorrow, for unselfishness and greed, for love and hate, etc.

In this number we first observe that potential duality which later on gives rise to all such contrasts, opposites and dualities, as good and evil, light and darkness, spirit and matter, and so on.

In the further consideration of this number two in its dualities and contrasts as good and evil which modern writers have clothed in a literal sense, giving a personality to them and calling them God and devil; this comes of continually looking upon the concrete side, and considering all things from a matter point of view.

It is well to keep in mind that while in the lower plane of life in its differentiated state, this number two is duality, but in the higher infinite realms it is in reality only potentially dual, that is, it is two in the one spirit matter, positive-negative, still is neither completely. The first duad in itself is as infinitely removed from mortal comprehension as is the naught or the monad, and the best that the mortal mind can do, is to consider its remote reflection in this world of matter under the same consideration as this number two.

The signification of this number in its primary effects is love, harmony and unity, although it may signify archetypally, separation or enmity. All religions or philosophies that are based on duality find their origin from this number two, not considering, however, the synthesis of two in one which it implies, and ignoring this unity of number one, which in its self-transformation forms the dual number, and which also lies back of the duality, and in reality reconciles all oppositions, no matter how irreconcilable they may seem to mankind. Both of these contrasted interpretations of this dual number are most aptly illustrated in the signification of the seventh house, which forms with the first house the first typical pair of opposites, and the seventh house signifies both love and enmity, also union and rivalry, though its primary interpretation is that union of two in one which, so far as the mortal mind is concerned, is found in marriage.

The monad may be represented by a dot or point; the duad by a line, and in this we have the point of the ascendant drawn out into a line and uniting the ascendant and descendant.

The fact is well demonstrated, that in the sign Libra the radical seventh is found the balance between the two signs Virgo and Scorpio, making duality of what should be unity, and unites Aries and the signs that follow it with the sign Scorpio and all that follow it, all representing the great evolutionary process, separation from unity, and back again to unity, but the mind can better comprehend the signification of the sign Libra as a balance when it is considered as the polar opposite of the first sign, Aries.

The sign Aries begins the circle and has the sign Libra to balance it at the opposite end of the diameter; in Libra we find the beginning of the southern signs and in Aries the beginning of the northern signs.

In the Zodiacal sign Taurus is number two, and this is found to be the negative side of Venus, as in Libra is found the positive expression, and if we will consider it as homogenious and imagine differentiation beginning in it, we will observe that we will naturally proceed from unity to duality, and the sign Taurus seems to imply this signification in its earthly nature, and suggests, so to speak, a central point with an environment, and manifests in reality the primordial spiritual source of that which in the manifested universe becomes cosmic substance or matter; the cause, so to speak, of illusion; and when all is unity the Deity gathers up into His consciousness all separated and differentiated modes of matter and consciousness in the universe over which he has rule, and the same with all, and thus synthesized in the Deity they exist as powers, vibrating in definite ways, He holding the centre unshaken in the very act of merging in or expanding into the infinite, the absolute, the superconscious, the one, and re-emerging from the latency of absoluteness at the dawn of the new universe.

On the re-emerging of the Deity from the latency of absoluteness at the dawn of a new cycle, or the beginning of a new world or universe, the tendencies to vibrate in these definite ways, as heretofore described, begin to be thrown outwards as actual vibrations, and thus in reality become the material basis of that universe, for matter as we have learned is merely a result of vibration and really consists of the memory of the old carried onward into the new, the legacy bequeathed by the old and is an inheritance to the new.

At this beginning of a new universe, the Deity draws in his consciousness from the depths of the absolute, and thus limits it to self-consciousness, and then, turning his attention to that which comprises that self-consciousness and its powers, start into activity, and then comes matter, the cause of all illusion.

The life breath goes forth, the Creative Deity, the centre of all is enveloped in matter, and sends forth his life-giving forces that permeate all where, and as that vibrating breath falls upon enveloping material and all is expressed in matter.

In this number two, we are only dealing with the potential root of matter, or that from which cosmic substance is created, actual matter comes under the higher numbers, and thus we have the dual force representing the powers or possession of the unified cosmic consciousness. In its symbology, this number is represented by the circle divided into two equal parts by a line drawn directly through the centre, and this divides the circle into two semi-circles, and the semi- or half circle represents the Moon in symbology and the Moon is found to be exalted in the sign Taurus, the second sign of the Zodiac. This is significant.

CHAPTER V

The Number Three is a Much More Mysterious Number Than the Number Two.

Now we may consider the figure three. With this number we have a trinity or what may be termed three in one, and much confusion has been created in the human mind in its effort to properly interpret this number. The dual number two being a two in one necesarily implies a triad or three in one, and is in itself a sort of potential triad. The number two is only transitional, is the mean between two extremes, and cannot exist by itself alone. It must either go onward into the triad or number three or backward into the monad, and then again, the triad being a three in one, therefore implies a tetrad or number four, and the process goes no further than this, for in the tetrad are to be found potentially all numbers, and thus we may find the mystic teachings contained in the very nature of the numbers themselves.

This number three is a much more mysterious number than the number two, that is, it is more difficult for the mortal mind to comprehend, though there is a vast amount of evidence of its existence as well as its operations in nature.

In first place, it might not seem a very easy matter to account for the origin of this number three, and if we fall back upon multiplication and illustrate the idea of the one becoming many, by the process of fission or self division in an organic cell, we have first of all the circle representing the cell, then follows the circle with the line across which gives the dual number two, and then this line is crossed by another line at right angles which takes us at the same time to number four, omitting the triad altogether. Then again, there is naught expressed in the triad that is not contained or at least implied by the duad; as the idea of two in one gives all that is required for the expression of the triad. This is demonstrated, if we fall back upon symbology of vibration. Imagine a point as the monad, which proceeds to vibrate backward and forward in a straight line. This line represents the duad and stands for all ideas of duality.

The idea of the two in one gives all that is required for the expression of number three, and this can be made clearer to the mind through the symbology of vibration itself.

Now we may consider a point which starts to vibrate backward and forward in a straight line, and keeping in mind the fact that this line represents the number two, and signifies all expressions of a dual nature, at the same time it is just as suitable for symbolizing the triad, as in the straight line there are two half circles and the centre, and we obtain no more than this from the number three, no matter what method is pursued as the two extremes, and the mean is all than can be obtained, no matter whether it be on the higher esoteric planes or in most concrete forms. However, we must remember that this straight line is a part of the circle. The point from which it begins is the absolute monad and the vibrations backward and forward are the upward and downward arcs of the circle, the in-breathing and out-breathing, the inhaling and exhaling, so to speak, and signifies the creation and dissolution. This centre between the two half circles, or the mean between the two extremes, gives the number three, but in the inhaling and exhaling there are really two such centres or means, that is, one at the beginning of each vibration, as the pause before exhaling, then pause, and then inhaling, and thus it goes on and on, this universal vibration, this inflow and outflow; but in a close observation we realize that we have number four and not three, as the number two multiplies itself into number four, really omitting the number three entirely, and a careful investigation reveals the fact that the first three numbers, in their own abstract region, are wholly beyond the understanding of the mortal mind in their proper interpretation, as in reality the first solid number, the first on any plane of concrete form, the first in the material manifested universe is the number four and back of this everything is abstract and informulate, so to speak, and not to be comprehended by mortal reasoning.

This number four gives form which we may observe by its powers and qualities, and these are firmly impressed upon it by the unveiled three, which are to mortal, abstractions rather than objects; at the same time it will be seen that they are the only causes of all things.

Thus in truth, all that the mortal mind can comprehend of the first three numbers is what may be realized of their effects and manifestations on the planes of concrete life. All the numbers are really contained in the decade and the three first numbers constitute the first three of the decade, and these are followed by the number four which stands for and represents material form, the egg of the universe, so to speak, and comparing to the fourth sign of the Zodiac, that is, Cancer; in this wise the primordial or first three are impressed as the manifested, or the secondary three, and the decade is therefore divided.

Thus three plus seven equals ten, and then the septenary is itself divisible, three plus four equals seven; so that the whole decade consists of three plus the the three plus four, or the abstract angle gives birth in the matter form to the manifested triangle, and this in still more concrete form is the creator or cause of the quaternary.

This number signifies the self-consciousness, the intellectual part in mankind and comprises object, subject, and the synthesis of the two, and that which was only potentially dual in number two is in process of further separation in this number three, so that we may say the number three comprises two extremes and their junction within itself.

The self-consciousness is really based upon the realization of the distinction between self and non-self, and the understanding of the principle of unity which underlies this duality is the great problem of the philosopher.

It is true the duality is clearly manifested in the sign Gemini, the third sign of the Zodiac, the twin sign and the first sign of a dual nature. This sign Gemini accords with a period of three years in mortal life, and intellect belongs here naturally for the reason that it depends upon a realization of similarity in difference, that is, the mean between the two extremes.

Metaphysical and Religious Trinities.

The number three is an expression of the first three numbers, and signifies a type of all metaphysical and the religious trinities, as in the trinity of the Hindus we have as the first person Shira, the transmuter, sometimes termed the destroyer, and represents the beginning and end of the universe. Then as the second personality we have Vishnu, the preserver, which harmoniously sustains all pairs of opposites and finally brings peace out of strife; and finally as the third personality of the trinity we have Brahmi, the Creator or Deity, working in the cosmic substance of things, and this process consists of energizing and evolving it, and refers to the universal process of evolution. This refers to the sign Gemini, and astrologically this process brings us in touch with the fourth sign of the Zodiac or Cancer, which sign, as we have heretofore inferred, stands for the egg of the universe, that is, the cosmic substance in which the Deity or Creator is manifesting, and in a metaphysical way these may be considered in the following order: First, Sat signifies being; second, Ananda, joy or bliss; and third, Chit, signifying intellect, as it will be found that all things in the manifested universe are classified under three heads according to the qualities impressed upon them by the action of the first primary triad in matter form, and from this are created another group of three: First, Tamas, signifying stability; second, Satva, signifies harmony; and third, Rajas, signifying activity.

These we have referred to heretofore, and in the consideration of these triad qualities the student will observe the similarity of these qualities to those manifested in the signs of the Zodiac, and are classified in this wise: First, fixed signs; second, common signs; and third, cardinal or movable signs.

We can give, without going into detail, a table that may be referred to containing a representation of all the metaphysical and religious trinities. This will prove helpful in assisting the student to a thorough understanding of this number three when used as a trinity.

First, in the Zodiacal trinity, we have the fixed signs as number one; common signs as number two; and cardinal or movable signs as number three. In the Zodiacal type of trinity, we have first, Leo; second, Virgo; third, the sign Libra.

In the Zoroastrian trinity, we have first, Churamazda; second, the twins; third, Armaiti.

In the Christian trinity, such as you are most familiar with in this country, first, the Father; second, the Son; and third, the Holy Ghost.

In the Orphic manifested trinity, we have first, Phanes; second, Urania; third, Saturn.

In the unmanifested Orphic, we have first, the Universal Good; second, Universal Soul; and third, the Universal Mind.

The human trinity is manifested first, in Atma; second, Buddah; and third, Manas. In the Buddhistic trinity, we find first, Amitabha; second, Avalokitishvara; and the third, as Manjusri. Then we have the Hebraic trinity, that is, first, Crown; second, Voice; and third, Wisdom. In the Egyptian trinity we have first, Ra; second, Osiris-Isis; and third, Harus. In the trinity as taught by Plato, we have first, Bound; second, Infinity; third, Mixed. In the Hindu trinity we find really three trinities, first Shiva; second, Ananda; and third, Brahma; and, metaphysically, these are manifested through the second trinity as, first Sat or being; second, Ananda or joy, pleasure; and third, Chit, or intellect; and by the action of the primordial trinity in their various qualities impressed upon these, we have the third trinity, that is, first, Tamas, or stability; second, Sattva, or harmony; and third, Rajas, or activity.

Thus we have these various trinities that have their origin in this number three, according to the ideas of the various religions and philosophies.

We have learned that the highest expression of the number one is the fixed, unchanging centre from which all things flow forth into universal manifestation, and to which all things return to be changed or transmuted, and it is really the highest extreme.

Then in the number two we find the mean of the two extremes, and taken together with number one we have the formation of the unmanifested triangle, and in this we may find the harmony, preservation and sustenance of all things. In number three we find manifestation and creation that is moving in universal cosmic substance, it forms the lower extreme and differentiates the tetrad.

Then, coming into the consideration of the qualities of the Zodiac, it may be readily observed how this number three, as containing the mean between the two extremes, links on to the astrological interpretation as given to the third sign or house, as messengers, methods of communication, journeys from one place to another, writings, etc., while this sign or house also represents the brothers and sisters. We find that the brother and sister have their unity or mean in the parent.

The fact of the triad or number three as signifying and implying duality is manifested quite clearly in Geometry, as the triangle is the first superficies and encloses space of two dimensions only, and the first proposition of Euclid is to construct a triangle upon a given straight line, and in order to accomplish this two circles cutting each other are used; and Helmis, and later Proctus, speaks of the signification of this upon the human soul.

CHAPTER VI

The Number Four is the First Number that Can Be Thoroughly Understood by the Mortal Mind.

We now come to the number four, the quaternary figure or tetrad.

As we have already learned, this number is the first solid number, the first number that can be thoroughly understood by the mortal mind, and the symbol of this number clearly expresses its interpretation in the concrete form, that is, the cross within the circle. The astrological meaning of this is the Earth as well as the part of fortune. The cross, as we have learned, stands for material manifestation, solidity and form, while on the other hand, the triangle stands in a general way for spirit formless, and it is quite evident that the one cannot exist without the other, for the triangle is in need of the cross to manifest itself, and the cross pre-supposes the triangle as its origin, and from which it was created. The whole decade is really contained potentially in the number four, as four plus three, plus two, plus one, equal ten. It will be seen that the numbers one, two and three are involved in the number four as is the child in the mother's womb, and while they are really earlier in their origin than the number four, they are, in fact, later than number four in their manifested reflections.

We may observe that the number four gives birth to the three first numbers, that is, through number four these first three numbers are enabled to find expression upon the lower concrete planes of life, and herein lies the mystery of ancient myths in which the male is the husband of his mother and so becomes his own father, that is, spirit, as the father represented by number three energizes matter, as the mother, represented by number four and impressed therein, is born as his own son, the lower expression of number three. The tetrad, or number four, is the number of the mother, and also of life in the matter forms, while in the astrological consideration we find that the monad, or number one, starts the positive male line of the horizon, while the duad, or number two, implies the vertical negative female line of matter, for the reason that it belongs to the triangle of the tenth house, though as it is not actually on that line, it merely indicates it as a potentiality in a similar manner as we have seen the dual number two indicating potential cosmic root matter, and the number three occupies a similar position, that is, with regard to the horizontal line as number two does to the vertical.

Now in the tetrad, or number four, we find a manifestation of both the horizontal and vertical lines as it is at one extremity of the vertical, and thus really implies the whole definite figure, in just exactly the same manner as it starts cosmically, definite manifestations in matter, and thus again we find the symbol of the cross within the circle is especially appropriate, not only from an astrological point of view but cosmically as well.

This cross within the circle was an ancient hieroglyph that signified lands, property, abiding place, etc., and this in its close association with the Moon and the fourth house may be readily observed. This number four in cosmic evolution is really the egg of the universe, or as we may say, the ocean of cosmic substance brooded over by the spirit, as the number three, the triad, the symbol being the triangle, stands for spirit, while the circle or egg is represented by number four and really contains within itself the seven-fold future universe, the central life and consciousness, which is appropriately typified in the Zodiacal sign Leo, which follows the fourth sign, Cancer.

In Leo we find the child manifesting from the egg, the god in the egg, so to speak, of Orphic and other throgonies, and we have already learned that the contents of the egg are sevenfold, are numerically the seven numbers which go to make up the decade, that is, cosmically the seven manifested planes with their divine rulers, or the seven divas, and Zodiacally the remaining signs of completing the circle.

This great ocean of undifferentiated spirit matter out of which a universe is to be created stands theologically for the divine maternal principle, that is, the mother god, and following after the trinity she gives birth on the lower planes of life to the second person thereof as her son, who is born of a virgin miraculously, that is, he clothes himself in primordial substance and starts it differentiating. This divine mother is quite closely associated in many religions with the incarnations or avatars of the second person of the trinity, and in connection with this fact we may observe that the Moon, whose home is found in the fourth sign, finds her exaltation in the sign Taurus, the second sign. The Moon, who has rule over the sign Cancer, we have learned, was represented by Isis, the divine mother.

In one aspect of the Egyptian religion, and in all great religions, there is always to be found first, the divine trinity, then the mother and her son, and then comes the seven, the son being the manifestation in matter of the second person and the highest of the seven.

We find the son the manifestation in matter of the second person, and the highest of the seven, and when the son has come into physical manifestation, he is then Leo, and conceived of the third aspect of the trinity, or, as we may say, the spiritual forces of the third Logos descending into virgin matter, evolve the entire universe, and to this stage, the sign Cancer, we may apparently relate the Goddess Nox as the mother and wife of Phanes in Orphic theogony, and here also is the crater or mixing cup containing the essence of the whole universe, and Proclus compares this crater with the egg, and Night the mother and wife of Phanes and Plato, in his psychogony, refers to the two mixtures or craters. In the one the Diety mixed the all soul of universal nature, and from the other was manifested the minds of men. The first of these is found in the sign Cancer, but further applications of a similar nature are found in the other two watery signs, Scorpio and Pisces, and in the sign Pisces we may find the dipper from which the water of the river of Lethe is drunk as the soul passes onward in its flight, and we may refer to the Orphic hymn to Night:

Hear, Night, parent Goddess, source of sweet repose,
From whom at first both gods and men arose;
Hear, blessed Venus, decked with starry light,
In sleep's deep silence dwelling upon night.

This term of the planet Venus applied to night reminds us again of the close relationship existing between the two signs, Taurus and Cancer, and which in ancient myths amounted to almost identity, also the relation previously referred to, between the second and fourth divine persons.

We may compare the qualities of the sign Taurus to either chaos or spirit matter, and out of it the egg which may be described as crystal white, is formed.

The cow was sacred to Isis, and Isis may be found often represented with cow's horns on her head. Plutarch says of Isis: "Isis is the female principle of nature and the receptacle in which all becoming or generation takes place." For this reason she is called nurse and all receiver by Plato, but by the majority, the one of numberless names, in that under the influence of reason or the Logos she receives all shapes and forms.

The affinity and close relationship that exists between the two signs Taurus and Cancer is clearly explained in

the cosmogonical significations of these two signs, as
Taurus stands for spiritual root matter, and Cancer for
the same substance that is formed into an egg, and under
the creative influence of the will of the Logos this stage
of the evolutionary process is aptly described in the
stanzas of Dzyan: "Darkness radiates light, and light
drops one solitary ray into the waters, into the mother
deep." The ray shoots through into the virgin egg. The
ray causes the eternal egg to thrill and drop the non-
eternal germ which condenses into the world egg, and the
three fall into the four while the radiant essence becomes
seven within and seven without.

In all solid figures the number four becomes the
tetrahedron or the three-sided pyramid, and this is the
first of the seven regular or platonic solids, the tetrahedron, the cube, the octahedron, the dodecahedron and
icosahedron, and these are all built up by combinations
of tetrahedra, two such forming the cube, and the octahedron and five of such forming the dodecahedron and
icosahedron, while the tetrahedron itself is made up of
two duads, each positive and negative, and united by triangular faces.

Kepler wrote concerning solids, "There are only five
regular solids, that is, it is only possible to make five
solids of different numbers of faces so that all the faces
of each solid shall be equal to each other."

The orbit of the Earth is the base of all and place
within this orbit, touching it at all points, an icosahedron,
and draw within it a circle that will touch all its faces
internally and the result will be the orbit of Venus; and
again, within the orbit of Venus place an octahedron and
draw a circle as before. This is the circle of Mercury.
Outside the orbit of the Earth place a dodecahedron,
Around this solid draw a circle. This is the orbit of
Mars.

In a further consideration of solids, recognizing the
fact that there are seven solids corresponding to the seven

planets, Neptune, and those planets beyond it cannot be counted as solids, for they are not in the true sense, as they have evolved into the condition of spirit matter, so to speak, though still possessing form. They are not composed of the same concrete substance as the Earth and other inferior planets, and the planet Urania is rapidly evolving to a similar condition through the evolutionary process that changes all things in the concrete expression, and which concrete form is lost altogether in time, as has been the case with Akassia, and for this reason it has not yet been discovered by mortal man on our planet.

We were considering the orbits of the planets and using the orbit of the Earth as a basis from which to calculate.

We have seen that by drawing a circle outside the Earth's orbit and inside this circle place a dodecahedron, and from this we find the orbit of Mars; then again, outside the orbit of Mars place a figure representing the tetrahedron, around this draw a circle, and this will give the orbit of Jupiter. Then again, if a hexahedron be placed outside the orbit of Jupiter and about this draw a circle so that the faces of the hexahedron will touch the circle internally and this will give the orbit of Saturn, and so on with the planets Urania, Neptune and Akassia. It will also be found that the observed distances of the planets from each other correspond exactly with the intervals between the solids.

This is significant and demonstrates the order that exists in the working out of the great infinite scheme of the universe, and each body is so placed in space that it will possess just the proper influence upon the rest, each possessing at the same time their proper qualities and distinct natures, in order to complete a grand harmonious whole.

We have found that the first four numbers, when added together, produce the decade, or number ten, which ends the first cycle of numbers, and potentially at least, includes all within its scope, and a careful observation will disclose the fact that this same condition is true in the universe if these four numbers are applied in a more extended manner.

We have learned that the number one is the absolute monad in the universe and number two is the second aspect of the divine trinity, which is, as we have found, potentially dual, and this stage provides us with the negatively existent or unmanifested triangle.

Then number three is the triangle manifested by the third person of the divine trinity and really includes the three higher planes of the cosmic seven planes, and the three Nirvanic planes, so called.

Then the number four, the tetrad, is the square or cross and represents the four lower planes or elements of the cosmic seven, and it is to these four planes that man is related when he is considered as a fourfold being, the four departments of his nature consisting, when numbered from the lowest concrete upwards, of, first, the physical man or activity; second, the astral man, or emotion; third, the mental man, or the intellect; and fourth, the spiritual man, or wisdom.

These correspond to the four triplicities of earth, water, fire and air, as their similar natures may be readily perceived. Septenary man evolves upon these four planes and in these four modes of his being.

The chief line of evolution now going on amongst the most advanced races of mankind, consisting of intellectual growth, though this side of man's nature is not yet fully evolved and will not be for some centuries to come. The spiritual side of man, the highest of the four, has scarcely been realized at all.

CHAPTER VII

The Number Five Has Many Meanings According to the Way in which it is Analyzed.

We will consider the number five, or what is termed the pentad.

This signifies five in one and is represented on a plane surface as the five-pointed star, and represented as a solid by the four-sided pyramid.

This number five really proceeds from unity for the reason that, similar to number three and number seven, it cannot be produced by multiplication of numbers, as we find that the number nine is the first odd number that can be produced or formed in this manner.

This number five has many interpretations and meanings according to the way in which it is analyzed, as well as the method of its application, for it may be considered either as four plus one or three plus two, and each of these vary again according as to whether the higher or the lower numbers are taken first. This is realized still more clearly if we consider its cosmic application with this number five which corresponds to the Zodiacal sign Leo.

We may begin to examine the contents of the egg of the universe. As we have learned, the whole of the universe is represented by the decade or number ten, though in its Zodiacal enumeration there are twelve, and of these we have already considered the first four stages, also the seventh and eighth, and in our investigation of the contents of the egg we find all the others from this period onward. We have really a double series of numbers, that is, one beginning with the absolute monad or number one,

corresponding to the sign Aries, and the other beginning
with that contained in the egg or that which is to manifest from the egg.

The sign Leo is number five when reckoned from the
first sign Aries, but when considered as the first and highest stage in the differentiation of the cosmic egg, it is
again number one in its expression, and we readily trace
its relationship with unity. The number one as the absolute monad in super cosmic silence and darkness, refers
to the first sign Aries, but when considered as the stage
of manifested expression and illumination in matter, it
refers to the sign Leo, and similarly the number five,
when reckoned from the first Aries, applies to sign
Leo, though when considered as the fifth plane of differentiation within the egg, we find refers to the tenth sign,
Capricorn, which we shall consider later when we arrive
at that point. It will not be difficult to understand that
this number five and its succeeding numbers possess very
different interpretations and applications, depending altogether upon the point or stage from which they are reckoned, that is, whether from monad Aries or the pentad
or Leo.

When seeking the highest application of number five
it may be found in its analysis into four plus one, that is,
we have come down into the matter expression, and in
order to proceed we must then return and take up the
absolute monad.

We have found the tetrad to be made up of four units,
the lowest of which is, potentially at least, several in one,
and at this stage of cosmic evolution with which we are
now dealing, this lowest unit stands for the whole sub-
stantial universe, which we may regard as one, but at this
stage differentiation begins within it, and we may sym-
bolize this by four circles, the lowest of which contains a
central point, and this gives very clearly the symbol of
the Sun for the pentad in its manifestation and close relationship to the sign Leo.

The sevenfold contents of the manifested universe are thus synthesized here in one unit, the highest of the seven, and the monad of manifestation, are just what Aries is as the monad in the silence of the unknown darkness to mortal minds though absolute light in itself, that Leo is as the manifested monad.

We have seen that the number one is represented in the Zodiac by the sign Aries, and number five by the sign Leo, and we find the Sun, which is representative of the physical manifestation of the Logos, is ruler and lord of the sign Leo, and has his exaltation in the first sign Aries. This great symbolism of the creative Deity being born from an egg is found more or less distinctively in various religions and myths. Among these are Chinese, Greek, Egyptian and Hindus. However, among the Egyptians the God in the egg of Orpheus or Ptah, as the active principle in opening the egg, stands for the first manifestation of creative light in the hitherto unknown darkness and mystery, and this is the stage of command, "Let there be light and there was light;" and Dzyan speaks in this wise:

"Bright Space, Son of Dark Space, who emerges from the depths of the great dark waters; He shines forth as the son, the white, brilliant son of the dark hidden father."

And from it is found these expressions that we may clearly understand just how all this symbolism is associated with the interpretation of the sign Leo and the fifth house; for here we have the birth of the child, which is the first definite expression, manifested, that is, bound by limitations, while at the same time these limitations are as boundless as universal space itself, comprising both its spiritual and material aspects; and this manifestation signifies the active representative of the creative Deity, that is, when it is clearly defined in terms of matter expression, and in him we live, move and have our being.

We can readily recognize the association of this ex-

pression with this fifth sign Leo, for through it is manifested energy of the Sun with his great life-giving forces,
though the individual is apt to use this force to evil as
well as to good, and supreme as the highest of the seven
cosmically, he becomes the type of the monarch of power
and majesty in the astrological interpretations of this sign
and house of Leo and the Sun, and when expressed in
terms of matter this is the highest of the seven cosmic
planes, and of these seven, the first three we have learned,
are classed as one and are called Nirvanic, and this is the
highest of them. Each one of these seven planes is under
the control of one of the seven Gods, as we have heretofore considered, and is related to one of the seven great
hierarchies, while in the Hindus cosmogony there apparently is no name or term applied to the highest of the
seven deities, as there are only five that are mentioned;
as the two highest are considered to be as yet concealed
from mortal mind, and not as yet revealed in this stage
of the world's evolution in which there is much truth;
though we find here Brahma the Demiurge, who is born
from the egg and makes all things manifest; and among
the Egyptians Seb was called the great cackler, as he was
supposed to have laid the egg from which the world
manifested.

It is well to refer to these things, as it shows that some
of the fundamental principles of truth had found expression with these various races in the past, though each
must necessarily clothe the thought in his own understanding.

The first act of creation began its physical expression
with the formation out of the primeval water, from which
came forth Ra, the immediate cause of all life upon Earth
in material form, at least, and the Almighty power of
this divine spirit embodied itself in the most brilliant form
and grandeur in the rising Sun, which stands as the physical manifested symbol. In the hymn that was chanted to
Amen Ra, or the Sun God of the Egyptians, he is ad-

dressed as "Thou Bull of thy mother," which indicates quite clearly the relationship of the self-produced son to his mother, as his own father, and similar to the relationship that exists between the two signs, Leo and Cancer.

In the Orphic theogony in the place of Brahma and Ra we find Phanes used as the manifestor, Protogonus, the first born, and called by Plato the animal itself, and the Orphic hymn to Protogonus begins in this wise,

"Oh, mighty first begotten, hear my prayer;
Twofold egg born, and wandering through the air,
Bull roarer, glorifying in thy golden wings,
From whom the race of Gods and mortals springs,"

and as has been said, the whole of this first and occult genera of the Gods which is called by the Chaldean theologists the intelligible triad or Leo, was represented by Orpheus under the symbol of an egg, on the exclusion of which by the Goddess Night, the God Phanes came forth, who is hence denominated Protogonus.

Phanes was symbolically represented as a dragon with the heads of a bull and lion, and in the midst, the face of a god with wings on the shoulders. The correspondence of this is quite evident, first, the Dragon, as Cancer, the Bull, Taurus and the Lion, Leo, out of which comes the God manifestation; although the interpretation of this symbol signified Pan, the all father or the universal creative power and the source of all life. This power was expressed by the golden wings and denominated Time and Hercules.

It was the Karmic ruler of the universe, for Necessity resides with him, which is the same as in the Nature law and which is extended throughout the universe, whose limits she holds in bonds of amicable conjunction, and this fourfold power corresponds to the Lipika of the stanza of Dzyan.

Thus Adrastia is the same as Nemisis, a daughter of

Night, Ocean being her father, and she was formerly regarded as a personification of the power which regulates and orders the spiritual and physical or matter world.

It will also be found that the judgment scene in the Theban edition of the Book of the Dead reveals the belief in the existence of a tri-formed monster, part crocodile, part lion and part hippopotamus, and whom the Egyptians termed Ammit, that is, the eater of the dead and who lived in Amenta. Her place is by the side of the scales wherein the heart is weighed, and it is clear that such hearts as failed to balance the feather of Maat were devoured by her. The crocodile was sacred to the Sebak, a form of horns. The Sun God and Sebak Ra is often depicted in human form with the head of a crocodile.

All these facts go to show the construction placed upon the interpretations of these numbers, and their close relationship with the Zodiacal signs. The fact that in Hindu astrology we find the fifth house or sign Leo is regarded as having to do with the past, Karma, makes these references to Adrastia Nemisis Ammit quite significant, and in this same system we find the crocodile again represented later on in the sign Capricorn, and wherein it acquires a human and personal significance.

The Sign Leo Stands for the Highest Manifested Power in All Philosophies and Religions.

We have already learned that the sign Leo stands for the highest manifested power, that is, God in all philosophies and religions, for whom the past, present and future are one eternal Now, as he is considered to be the great controller and guide, energizing and directing all things. He stands for the manifested expression of the unmanifested one, and there is no life apart from his life in the whole universe. Thus he is represented as the supreme ruler of Karma.

Phanes was termed the ruler of love or eros. This is represented by that primal love or desire or Karma Deva which sprang from the all, and in the words of the Rig Veda, the primal germ of mind, while in the Orphic hymn he is termed Priapus, a reference to the same creative desire in the universe, though in later times the term has become grossly Phallic, and the application of this to the sign Leo is seen at once. The desire which first arose in it is first seen in the martial Aries, but is manifested in Leo.

We have seen how the desire that was created in the monad or sign Aries comes to active expression in the matter plane of life in number five or tetrad or the sign Leo, and we may find the first of the seven great hierarchies represented by this sign.

The hierarchy of the creative powers is divided esoterically into seven, that is, the trinity of three and the four manifesting in physical life within the twelve great orders recorded in the twelve signs of the Zodiac, and the highest group is composed of the divine flames, so termed, and which are also spoken of as the fiery lions and the lions of life, whose esoterical interpretation is contained in this Zodiacal sign Leo. Here, too, is found the hierarchies of the Dhyani Buddhas. Their state is that of Parasamadhi of the Dharmakaya.

The meaning of this state or condition is non-progress, and where progress is not possible, the entities existing there are said to be crystallized in purity in homogenity. The hierarchy of non-substantial primordial beings is a place or state that is really no state at all. This hierarchy contains the primordial plane, that is, all that has been, all that is, and all that shall be from the beginning to the end of the mahamanvantara; all is there, etc., and this is based upon the higher manifestations of this sign Leo.

The signification of the tree in symbology is familiar, as it represents the much differentiated universe, and this figure representing the tree is drawn with its roots above, reaching into the higher spheres and planes, while its branches are bending downward in the lower planes, and this is significant of the fact that visible matter is rooted in the invisible spiritual.

Then, if we take the four-sided pyramid which is representation of the number five in its solid matter form, and apply it in the same manner as the tree is applied, we shall treat it like the tree, that is, its four square base will be reaching up above into higher planes, and represents the sacred tetraktys, while the apex below will signify the manifested light or fire of this sign Leo, and will then correspond to the analysis of the number five into four plus one in this order. However, if the order of these two figures is reversed, that is, one plus four, and the pyramid is represented with its apex reaching upwards, the signification would then be entirely changed, for the base represents the four lower or cosmic planes of manifested life in the concrete, and apex represents the three higher planes, which may be regarded as one in their manifestations so far as the lower planes are concerned, though the pyramid includes all the numbers up to the pentad, for its apex signifies unity, its sides are triangular and its base quadrangular.

Now the intelligible world proceeds out of the divine mind in this wise, the tetraktys reflecting upon its own essence, that is, the first unit, the producer of all others on its own beginning, thus once one, twice two, and immediately is produced the tetrad or number four, having on its top the highest unit, and becomes a pyramid whose base is a plain tetrad answerable to all superficies upon which the radiant light of the divine mind produces the form of incorporeal fire, by reason of the descent of Juno, known as matter, to inferior beings. Thus comes essential light, not burning but illuminating, and this is the beginning of the middle world which the Hebrews term the Supreme or the World of the Deity.

We will proceed with our subject concerning this essential and illuminating light that was termed the supreme world of the Deity by the Hebrews. It was also termed Olympus, was entirely light, and replete with separate forms, where is the seat of the immortal Gods, whose top or apex is unity, its walls trinity, and its superficies quaternary. Olympus or the intelligible world is the manifested triangle of the three highest cosmic planes, and find a similarity in their interpretation to the signs of the Zodiac, Leo, Virgo and Libra.

Thus the division of the number five, the pentad, into three plus two is most aptly illustrated in the application of this number to the sign Capricorn, and upon careful investigation it will be found to have a close correspondence to this sign, but is directly associated with the fifth house affairs, as may be readily seen.

The ancients in various nations considered this number five to be a symbol of marriage and generation as well, for the reason that it includes the first odd or masculine, also the first even or feminine numbers, that is, three plus two equal five, and this number and term was applied to the planet Venus, also Cytherea, Lucina, Juga, Opigera, and included other deities who presided over nuptials and parturition. At weddings the Romans had a great regard for this number five, and as a practical display of its reference to the marriage, five wax tapers were always lighted and placed in some conspicuous place as a symbol that could not be misunderstood. The Romans were also accustomed to use this number freely in matters of divination as to whether good or evil fortune was to come to the newly wedded couple, and Plato makes mention that he would have admitted the wedding guests by fives, and Jesus in the parable of the marriage spoke of the bridesmaids in fives.

It is quite remarkable how widespread these ideas associated with the duad or number two and the triad or number three as composing this number five; for instance, the second sign of the Zodiac, Taurus, is ruled by the planet Venus, which in Greek aphrodite is the typical woman; while the third sign Gemini is ruled by Mercury, termed Hermes, the typical male, and the combination of these two gives the word hermaphrodite, that is, implying both male and female, and referring again to Hindu astrology* of the periods and sub-periods, it will be observed, that whenever Venus and Mercury combine by their period and sub-period, something concerning marriage, love, affairs, the wife, children or pleasure will manifest itself, as for instance in the Venus sub-period of the Mercury period, if other aspects concur the native will find pleasure through the wife and children at such times.

The principal point to be considered and to which the attention is called most particularly is not so much that the combination of Venus and Mercury is favorable for marriage, etc., but that this influence is especially favorable for children, and such affairs as we have learned relate directly to the fifth house, the house ruled by Leo, and which we have learned has direct rule over the affairs of the offspring, and in a horoscopal figure when these two planets are combined, that is, Venus and Mercury, the native will have rule and power over his fellowman and rise to such position of authority through this influence if other influences accord. This refers directly to this number five as exemplified in both Leo and Capricorn. Both are signs of power and majesty.

The numbers similar to the signs of the Zodiac all have their special significance and importance in the great universal scheme.

Each is essential to the manifestations of the other, in order to have a perfected whole, all working in the utmost harmony with the planetary system. Though the results may not always seem to bring harmony, still

*See Vol. IV., Page 72.

it is so, harmony with the purpose of infinite love and wisdom.

The two Zodiacal signs, Gemini and Taurus, the third and second signs that go to make up the number five, manifest a peculiar relationship to the signs Cancer and Leo. Cancer is the second sign from Gemini, and the third sign from Taurus, or if we will substitute numbers, it is the second from the third and the third from the second, and the duad and triad meet here in a kind of mutual reversal, which fact gives them particular significance. The combined influence of the two signs, Taurus and Gemini, is cast in germ into Cancer, and exists in the tetrad embryonically, similar to the child in the mother's womb, and it is subsequently born into expression as is the completed pentad in the sign Leo, The Child. This has a cosmic significance. In this wise, brought into manifestation by Brahma or the third Logos, the second, the generating father, Vishnu, places that germ of life in order that it may unfold and develop therein.

The Zodiacal sign Cancer represents the undifferentiated non-atomic matter in which the vivified germ is placed, and Leo stands for the highest manifestation of the universe that is born from the germ. Thus by close study it will be clear that Leo, corresponding to the number five, has a remarkable bearing in its assistance from unmanifested to the manifested.

CHAPTER VIII

The Number Six a Manifested Duad Pure and Simple.

We now come to number six, and this number is represented or expressed on a plane surface by the two interlaced triangles forming the six-pointed star. In the solid figure it is represented by the cube. It is quite evident that the higher the number the more difficult is its analysis and the more complicated does it become.

We can regard this number six as three plus three, or four plus two, or five plus one, though all these analyses carry with them a strong suggestion from the duad from which this hexad or number six is created, and to which number it will be found to possess a most remarkable resemblance. The duad, or number two, is the second number within the cosmic egg, four plus two, and the corresponding signs, Taurus and Virgo, belong to the same earthly triplicity.

Then again, the correspondence is shown being the sixth number it bears a strong relation to the sixth division of the cosmic egg, and which will be found to be represented by the Zodiacal sign Aquarius, the eleventh sign, and this number eleven again suggests duality in this wise, the ruler of the sign Virgo is Mercury, and this planet finds its exaltation in the sign Aquarius, and is therefore strong in this sign, taking the number six or hexad as the second number or plane from the cosmic egg itself, it is really a manifested duad, pure and simple, as four plus two, being the second or middle number of the manifested triangle formed by the Zodiacal signs, Leo, Virgo and Libra, as can be clearly shown.

If the Moon be taken as number four and Venus as

number two, the following influences in Hindu astrology manifest during the Venus sub-period of the Moon's period is of interest in the consideration of the hexad or number six, as at such times there will be sickness and suffering from pain, bodily affliction, enmity, loss of wealth through enemies. From these influences can be clearly noted the relationship that exists between this result and the sixth house or sign, Virgo, the house of sickness and bodily afflictions, and its opposite the twelfth, the house of secret enemies, and the system of Hindu* astrology recognizes the correspondence to some extent.

The sign Virgo, in its character of the manifested duad, corresponds to the second hypastasis of the triple atina as well as to Buddha, and also to the second concealed God of the seven to the second Nirvanic plane, Paranervana, and to the second great hierarchy, and of this second great hierarchy we may consider in this wise, possessing their potential being in the higher group they then become distinct and separate entities, and these are termed the Virgins of life, the great illusion, etc., and when taken collectively are known as the six-pointed star, the double triangle.

We have learned that the first born of life are the heart and pulse, so to speak, of the universe, while the second constitutes the mind or consciousness of being; and here are the hierarchies of the celestial Buddhas, or Bodhisattras, who are said to emanate from the seven Dhyani Buddhas. Then, too, in referring to the abode of Vairajas, the Pitri Divas of the Sun, these belong to the fiery egos of other manvantaras. They have already become purified in the fire of passions and it is they who refuse to create. They have attained to the seventh portal and refused Nirvana, remaining for succeeding manvantaras.

We may refer to these terms and oriental teachings

---

*See Hindu Astrology, Vol. IV.

and philosophy in order that the connection may be more
clearly seen and that some basis of truth exists. It also enables the student to better understand some of the higher
interpretations of the sixth sign, Virgo, which has rule
over the sixth house. As those who refuse Nirvana are
classed as the servants of the Great One, they have
reached the highest portal leading into Libra but have not
yet entered in.

The double triangle, or the six-pointed star, is also
associated with Vishnu in India, and this again takes us
back to the duad, the second hypostasis of the manifested
trinity.

The star itself is of dual nature, being composed of
two triangles which, when taken as contrasted wholes,
embody very similar ideas as those that are expressed
in the duad or number two, only in greater detail, and
as a symbol this Solomon's seal, as it has been termed,
is of the greatest significance though it resembles the
duad in not being complete in its being, impossible to
dwell upon it.

Almost every meaning the symbol contains turns upon
the presence of the central point in the two triangles,
and this in turn converts it into a significator of the
heptad, or number seven.

Thus it will be found that the six forces which are
referred to the sign Kanya, meaning Virgo, the sixth
sign, are really incomplete without the synthetic seventh.
Then again, the double triangle or six-pointed star, refers to the six forces or powers of Nature; the six planes,
principles, etc., all synthesized by the seventh or the
central point in the star. All these upper and lower hierarchies included emanate from the heavenly or celestial
virgin, the great mother of all religions, the Androgyne,
the Sephira, Adam, Kadmon.

In referring to the Virgin, the great mother in all
philosophies and religions and the same that is given
to the sign Virgo, the sixth sign of the Zodiac, it will

be found that this Virgin, when considered from the esoteric point of view, really belongs to the sign Cancer more than it does to the sign Virgo.

A Hierarchy is a Congerie of Otherwise Separate Intelligences Voluntarily Associated Together to Carry Out One Common End.

We will now consider just what composes these six hierarchies, also these six forces in Nature to which we have heretofore referred, and the manner in which they are all so closely connected with the sixth sign of the Zodiac, Virgo.

A hierarchy is a congerie of otherwise separate intelligences associated together, nor arbitrarily or compulsorily, but voluntarily, and in accordance with natural evolution to carry out one common end.

The analogy of the physical body itself affords a very good illustration of this fact, as it is composed of a multitude of cells each with powers and activities of its own, and each with its own special limited form of consciousness; still the whole are combined and co-ordained together to sub-serve the purpose of one central consciousness greater than all, and which may be termed biune or animal or triune, the human monad, the soul, which utilizes them as a vehicle for such a period and time as they can co-here together into one great whole, and in a similar way is the hierarchy. It is composed of a vast number of highly evolved, self-conscious intelligences who have, without losing their individuality, merged themselves together to form one vast vehicle which shall be energized and guided by the one great universal central consciousness, even more grand and sublime than the total of their own. This great universal central consciousness we term the God, and in the concrete expression, the hierarchy is the wife, the power or shakti, and from this fact arises the six Gods and their six consorts in various philosophies, religions and myths. Thus the Gods are the directing will and guiding spirit, so to speak, of the whole, while the hierarchy is the executive, acting as the agents or servants whose mission it is to carry out his will on the lower planes of life.

It is true that every living soul in the universe, whether human, super-human or sub-human, is slowly but surely evolving under the power or influence of one or another of these great groups, and it is destined to conscious, sooner or later, with that to which it rightfully belongs as a lawful heir.

In considering the Zodiacal sign Virgo, we may discover not only separate souls combining into hierarchies, but even much more than that, as we find six otherwise separate hierarchies united into one whole, to sub-serve the will of the one great One, which we find expressed in the sign Leo, and these then truly comprise the six forces of nature that are signified by the sign Virgo, and they are really the servants and agents, or we may say, the executives of the royal ruler as exemplified by the sign Leo, while the significance of service is found in the sixth sign and house of Virgo, and this makes quite plain and clear the meanings of this sign and house as interpreted by the science of astrology, and it is quite significant that one of these interpretations, that is, health and disease, which is otherwise somewhat mysterious, is herein accounted for. And carrying the analogy down to the physical body of man, the seven hierarchies become seven nerve plexuses, the centers and physical vehicles of the astral.

In considering the similarity of the seven hierarchies as compared with the physical body of man and woman, we see how they correspond to the seven nerve centres and act as the vehicles for expression of the astral, and the lower six of these are considered as one in the sign Virgo.

The sign Leo represents the vitality which actuates and co-ordinates them, and upon their harmonious action and co-operation the health and life in the physical form depends, and from this we may see how a malefic aspect to the conditions of health may bring ill health, though the primary interpretation of this sign and house, as indeed of all the twelve, are good, and it is only when they become afflicted by these malefic aspects that the evil manifests itself.

There are none of the Zodiacal signs, houses or planets that are essentially malefic; while everything is a result of harmony or inharmony, unity or opposition.

We have learned heretofore that just exactly as the sign Capricorn is associated with the sign Leo, in the same manner is the sign Aquarius associated with Virgo, the sixth sign, for as we may see, the sign Aquarius represents the sixth element in the egg. We have learned that the sign Aquarius and the house it rules, the eleventh, has to do with the friends of the native, and really signifies the people associated together for one common purpose. Thus we may say, the senate may be said to come under the rule of this eleventh house in mundane astrology, for in a way it is the vehicle or executive agent of the President and thus its meaning of societies, associations, companies, etc.

The analogy of this with the hierarchies and shaktis, or powers of Virgo, is too obvious to pass unobserved, and a careful investigation will demonstrate just how closely all the houses and signs are associated in their separate or collective interpretations; and then too, the sixth house signifies the army and navy, or what we may consider the executives or protectors of the common people and the nation.

A Synthetic Influence Manifested by Urania.

The planet, Urania, we have learned, has rule over the sign Aquarius and the eleventh house affairs, which a careful study of its nature will show to be true, though whether its nature is more allied to a co-operative hierarchy than the in-dwelling and directing will, may not seem so clear to the mind, but in this, as with other influences, we must consider whether it is manifesting upon the higher planes of life above, or on the lower planes of life nearer to mortal expression.

We may also find that Mercury is closely associated with this airy sign Aquarius, as we have already learned that Urania is a higher Mercury Venus in its expression; and, in fact, Mercury was the diurnal ruler of Aquarius and the nocturnal ruler of the sign Virgo, while Urania was the nocturnal ruler of the sign Aquarius and the diurnal ruler of the sign Virgo.

This is a subject of great interest to the student in making his calculations from the eleventh and sixth house affairs.

There certainly is a synthetic influence manifested by Urania and it is significant that the period of this planet, that is, eighty-four years, is the product of the twelve signs of the Zodiac multiplied by seven, the number of planets.

In the Orphic system, the sign Virgo refers to the noetic-noeric triad, which is really part noetic and part noreic, but wholly neither, in the same manner that Virgo is a common sign, though it is neither fixed nor movable in its nature, but is really an intermediate, so to speak, between the two extremes.

In considering Urania in the Orphic system, we find Orpheus placed the God Urania in this order, "ethereal earthly, whose all various frame azure and full of forms no power can tame, all seeing source of Saturn and of time," etc., and in the consideration of these Orphic hymns that were taken from very ancient ideas of wisdom, we learn that every intellect either abides, and is then intelligible as being, better than motion, or it is moved and is then intellectual; or it is both and is then intelligible and at the same time intellectual.

The first of these is Phanes; the second, which is alone moved, is Saturn; and the third, which is both moved and permanent, is heaven; and thus we have again the fixed, movable and common influences, the latter being that of Virgo and Urania, thus bringing out anew the dualistic nature of the sign and God.

We may, if we desire, classify these in a manner that will be quite intelligible, as for instance we take the numbers. We have first five or one, second six or two, and third and last seven or three. Then, taking the signs in their correspondence to these numbers, and we have the sign Leo, corresponding to number five as the fifth sign from the monad, and also to number one as the first manifestation from the egg represented by the sign Cancer. Then, second, we have the sign Virgo corresponding to number six, being the sixth sign and number two from the egg, and third we have the sign Libra, corresponding to the number seven, and three being the seventh sign from the monad and the third sign, the triad or triangle from the egg, whose combined powers contains the whole decade itself, as we have learned.

Then as regards their quality. We have first number five and one corresponding with the sign Leo, and find the quality is of a fixed nature, Leo being one of the fixed signs in its Zodiacal influence. Second, in referring to the number six and two and the sign to which they correspond, we find the quality is neither fixed nor movable, but an intermediate between the two, and is termed common in its Zodiacal nature. Third, in the number seven and three we find their correspondence in the Zodiacal sign of Libra, and the quality we find is movable in its Zodiacal influence. Then, referring again to the Orphic system, we find a God or power back of each, as for instance, corresponding to numbers five and one, and the sign Leo and fixed quality, we find Phanes as the God, and Nox as the Goddess; second, in correspondence to the numbers six and two, and sign Virgo and intermediate, or common to both quality, we find the God is Urania, the Goddess is named Goea; the third and last, in correspondence to number seven and three, and the sign Libra, with the quality movable, we find the God represented or power back of this sign to be Saturn and the Goddess Rhea.

A careful study of these classifications will be of great importance in obtaining a thorough understanding of interpretations on the higher planes of expression, and the esoteric meaning that is not implied in an exoteric view, as given when only considering their exoteric value or their application to the physical planes of life, and we refer to these numbers, signs, qualities, etc., for the reason that it brings out more vividly their influence that is naturally associated with the individual at the time or moment of conception, and may be studied with benefit in the judgment and influences as manifested at that time.

It will be found that the six forces as signified by Kanya, and to which we have heretofore referred, are in their combined influence represented by astral light. This does not refer specially to the astral planes or sphere unless it may be by downward reflection, so to speak, though the sign Aquarius, and Virgo will also be found to represent the shakti or powers termed the female energy of the Gods and the crown of the astral light.

We find Buddha was considered the God of the planet Mercury, which planet has rule over the sixth house and Zodiacal sign, Virgo, and this astral light is quite similar with Goea the Goddess, the shakti, or powers and executors of the God Uranus, and which we may consider as an aspect of mother Nature as is found expressed in this sixth sign Virgo.

The hexad in ancient times was termed Kosmos, thus showing a similarity or relationship to that element known as the cosmos, of which we have learned somewhat in our former considerations of the primary numbers.

It may be well at this time to give a word of caution concerning the name of the Greek God, Uranus, having been applied to a planet recently discovered, or rather rediscovered, and sometimes known as Herschel. This planet was originally named Uranus, and from this comes the name Urania or Uranus, for the reason that the God Uranus was considered to find a physical manifestation through this planet, and a further investigation as to the true occult and mystical nature as well as the evolutionary process of this planet and its elemental forces in manifestations, will demonstrate this fact beyond a doubt, and make clear to the present race the influence as manifested upon our planet the Earth, and there is one part of the nature we may consider at this time, and that is that while the God Uranus and the Goddess, his especial consort, are associated with the sign Virgo, this planet is really, so far as its influence upon this Earth planet, is of a nervous, mental and Mercurial nature, and herein we find the close relationship between the planet Mercury and Uranus, and herein also is found the difficulty to be confronted with the individual who has the planet Urania prominently configurated in their horoscopal chart of birth. The native receives these higher vibrations of Urania, and when he is himself largely environed in the earthly triplicity, the vibratory rays, as they reach out to penetrate through this matter substance, are so colored by it that the true Uranian principle does not find expression whatever, and there is a constant contradiction going on. The native is mystified, he does not know which way to turn;

his mental faculties are at a loss to act properly and the soul itself is lost sight of, though it is seeking to obtain a hearing, but the environments are too thickly permeated with the lower material and the higher spiritual senses are dulled. They awaken at times to the call, but lapse again into unconsciousness before they comprehend its import, and thus it goes on until there is a final awakening.

This number six, or the hexad, was also held to signify the health separate and apart from the astrological associations, that is, in its physical aspect, and was termed Panacea, a familiar term that has been handed down from that time, and the number five also had to do with the health, but this was due largely to the vital energy implied by it, as we have learned.

The number six we find is symbolized by two triangles and was considered to belong to the planet Venus, for the reason that those may stand for the union of the sexes. The Zodiac we find is twelvefold, while numbers are only tenfold. If a cycle of ten be halved, the first part of the second half will be number six and the first part of the second half of the Zodiac is the sign Libra, the home of Venus, corresponding to the descendant of the seventh mundane house, marriage, and it will be observed that the same planet Mercury is associated with Gemini, the triangle, and Virgo, the two triangles.

In the Orphic hymn to the earth we find an apt illustration of the relationship between this number six and the season of the Sun in the sign Virgo; as the number six was considered to be the symbol of the Earth during the autumn and winter sleeping months, and the number seven during the spring and summer months. The symbol as the spirit of life animated her at that time, the central force, or the number six, was the symbol of the Earth planet, ready to be animated by the divine spirit. and which influence we may aptly compare with the relation as exists between the signs Leo and Virgo. The

Orphic hymn may be given here in its reference to these influences:

All parent bounding whose prolific powers
Produce a store of beauteous fruits and flowers,
All various maid, the immortal world's strong base,
Eternal, blessed, crowned with every grace,
From whose wide womb as from the endless root,
Fruits many formed, mature and graceful shoot;
Deep bosomed, blessed, pleased with grassy plains,
Sweet to the smell and with prolific rains, etc.

Then, again, we find a similar expression in the ode to autumn, as follows:

Season of mists and mellow fruitfulness,
  Close bosomed friend of the maturing Sun,
Conspiring with him how to load and bless
  With fruit the vines that round the thatch eves run.

And further in the Orphic hymn we may read, "O Mother Earth, of Gods and men the source," and as we have seen, Hermes or Mercury is born of Maia, mother of Hermes and still youthful Maia, and we find Maya to be the mother of Guatama Buddha, and the Virgin Mary as the mother of the Christ.

There Are Many World Saviors Allegorized as Having Been Born of Virgin Mothers.

There are many world saviors allegorized as having been born of virgin mothers prior to the birth of the Nazarene, and for instance we can realize that when the sign Virgo rises or is on the ascendant, having rule over the first house, then Capricorn, representing the human ego, will occupy the fifth house and thus stands for the offspring of the virgin, considering also that the fifth house is the house of children, and therefore it is this ego that must be crucified and descending into matter, must be tempted by Satan or Saturn and thus undergo initiation by rising into heaven, or otherwise into that higher state of conscious being which we find signified by the planet Jupiter.

Then, again, concerning the number six the pagans considered as Bereschith, of which they receive their Berecynthia, or the mother nature of the gods. The term Bereschith signifies Genesis, the great source, the great head from which the beginning was, and if it be used as a name it would then be feminine, and then the singular Bara, which follows it, will no longer seem strange and it would then be translated in this wise, the eternal Genesis created the Gods as the ternary expresses an absolute and complete concept. Two ternaries give the idea of two absolutes and under this aspect the six is only the binary exalted and carried to its supreme power.

In this way it can figure the revolt of Satan or the spontaneous deification of the created spirit, a conception that is quite absurd though sublime to a high degree. It is absurd for the reason that two infinities are not possible, and still it is sublime for the reason that this impossible is in some manner realizable in the indefinite extension of liberty. This idea of revolt seems to manifest quite naturally under the heptad, but we find that in our former consideration of the number seven and eight, the heptad and ogdvad, that these numbers were closely associated together with the hexad, or number six.

We find the numbers six and eight closely associated, as we do their representative signs Virgo and Scorpio, the sixth and eighth signs respectively. In the sixth house we find reference to disease and ill health, bodily afflictions, etc., while with the eighth we find the significations of disintegration of the physical body and which is termed death to the physical form. Then, too, we have seen the association of the planet Urania with Virgo, the sixth sign, and the occult significations of Scorpio are well known, in fact we find the results that begin their expression comparatively in the sixth sign, Virgo, find their full manifestation in the sign Scorpio, the eighth sign.

Concerning the duad, a word here may be well. In the interpretation of this number in its general significance, it is made to express both evil and opposition, but in its higher significance the fact must not be lost sight of, that it signifies ideal love and unity, and it will be seen in Nature's law that opposites attract one another, as may be exemplified in positive and negative electricity. We have already learned that separation begins with the duad, though it is really only potential in its manifestation and before it is really separation the triad is necessary. The pair of opposites would be incomprehensible without the tacit assumption of a synthetic third stage.

The universe is pervaded by three qualities and not by two alone. Number one is absolute unity; number two is potential duality; while number three is the actual duality and as heretofore shown in the table given in any triad one number is positive, one is negative and one is common to both or intermediate to both, or as we have observed in Zodiacal terms, one is fixed, one is movable and one is common to both in its expression as manifested by Leo, Libra and the common sign Virgo.

It is a fact that in order to have complete opposition there is a previous unity implied, and this is demonstrated by the signification of the triad, or third number of the triad, which is found to be allied to both opposites and forms the connecting link, and it is significant that the third sign of the Zodiac is the first one that is dual in its nature, and the sign Libra, which is the third sign within the egg or sign Cancer, has opposition, war and combat as one of its significations, and has also the seventh mundane house, the descendent.

The nature of the sign Aries and its ruler Mars, the opposite of Libra, is well known, and while the symbolism of the fall of the angels and the war in heaven find some significations in the signs Virgo and Scorpio, it really comes under the sign Libra, occupying as it does the sign upon the descendant, and it is well known that a planet found here is not so powerful in its nature in the horoscopal figure, and the influence for good or evil is lessened.

In Hindu astrology one of the principal meanings given to the sixth house is enemies and the general interpretation gives to the descendant the significance of opponents.

Considering number six as composed of two binaries, we may say six is the number of mankind. It is the number of struggle, of love and liberty, and it also represents the woman clothed with the Sun and having the Moon under her feet, who cries out in the pangs of childbirth.

The idea of labor and servitude arises out of the passivity of Virgo following after the superior will of Leo. The ideas of love and of man seem to associate rather with the septenary, for this provides the central point which separates, and at the same time unites the two triangles, and under the empire of the septenary it is both grace and love, though it is the septenary that both rebels and loves.

CHAPTER IX

The Number Eight is the First Cube.

The number eight is capable of a variety of subdivisions and applications. It is, in fact, the first evenly even number. It is the first cube. It very aptly illustrates the laws of inversion and reflection and it contains potentially the whole decade within itself.

In using the term, an evenly even number, it is meant one that may be halved and these halves again halved without an odd number making its appearance and in this wise going on back to unity itself. They are formed by beginning at unity and continually doubling it.

We have learned heretofore in considering the sex manifested by the signs of the Zodiac, that the even signs are female and the odd signs male, and this also applies to numbers themselves, that is, the evenly even series is pre-eminently female, since there is no odd number that appears therein after leaving unity, and the Deity being considered as one, we can readily see in modern times how these individuals considered the One as God, and one being considered masculine, although that this was not the original conception of those who mastered these truths in past ages, is quite evident.

We find authority in ancient writings concerning female deities, and the peculiarity of the female character as manifested by the number eight was the reason it was associated with female deities and more particularly with Rhea, the wife of Saturn, and the mother of the manifested Gods, as Saturn was considered the father according to ancient mythology, and considering the eighth sign of the Zodiac, that is, Scorpio, and it was for this reason that circumcision was performed on the eighth day and conception also comes under this number.

It will be observed that the other signs stand in peculiar relation to Scorpio when taken in pairs according to their planetary rulers, as according to the ancients, each one of the seven planets rules two signs, one feminine and one masculine; that is, Mars ruler of Aries, masculine, and Scorpio, feminine; Venus of Libra, masculine, Taurus, feminine; and so on with all the planets and signs of the Zodiac, the luminaries, the Sun and Moon being considered together as one in rulership over Cancer, feminine, and Leo, masculine.

In a further consideration of the number eight and the peculiar relationship that exists when taking the signs in pairs according to this rulership, we find in the first instance that Mars rules the sign Aries, the first sign of the Zodiac and Scorpio the eighth sign. Now Scorpio is found to be the eighth sign and the eighth from the first. Then Venus has rule over Taurus, the second house, and Libra, the seventh house, and here we find Scorpio second from the seventh house, and seventh from the second. Then Mercury rules Gemini, the third house, and Virgo the sixth house, and again we find Scorpio the sixth from the third and the third from the sixth house. Then, again, Jupiter has rule over Sagittarius the ninth house, and Pisces the twelfth house, and again we find Scorpio is ninth sign from the twelfth and the twelfth from the ninth house, or sign.

Then Saturn has rule over Capricorn the tenth sign, and Aquarius the eleventh sign, and we find Scorpio is the tenth sign from the eleventh and the eleventh sign from the tenth. Finally the Sun has rule over Leo, the fifth sign, and the Moon over Cancer, the fourth sign, and Scorpio is the fifth sign from the fourth house radical, and the fourth sign from the fifth house radical. This shows the relationship to Scorpio that exists with none other of the signs, showing this sign to be of the greatest importance in the consideration of the horoscopal figure. It is one of the most fruitful signs, being female in nature.

This peculiar reflection and reversal of the pairs of opposite or twin signs does not exist in any of the signs except Scorpio, and is certainly characteristic of the power of this sign in the material consideration.

This is found in conception and death of the physical body, so called, as the sign Scorpio rules the eighth house, the house of death, and these two events of the ego's conception into matter expression and transition from the same, come under the domination of the sign Scorpio, and we observe its significance in its effect of transference from one plane of life to another.

Then, again, if we may consider the Sun and the Moon as the parents, we observe that they meet as they enter the sign Scorpio, and under this peculiar reversal conception takes place, and then allowing nine complete signs to elapse as they come in the regular order of the Zodiac, we come first to Cancer, the eighth from the sign Scorpio, the great mother sign with child, to Leo, the ninth from the eighth, that is, Scorpio; and Leo, we know, is recognized as the fifth house, the house of children, and thus the Sun, which stands for the procreating father in the sign Scorpio, is manifesting in Leo his own sign.

In the ancient astrology, this Earth planet was regarded as being surrounded by seven typical planets, that is, the sphere of the Moon coming first and finally Saturn being the farthest away, and the eighth sphere or plane was that of the fixed stars from which the ego came to Earth, and it thus synthesized the whole seven.

In considering the sign Scorpio, the eighth or ogdad, and its relation to the Zodiac, we observe that Scorpio forms the second tetrakys and repeats in a secondary stage the process as found in Cancer.

Cancer is a synthetic sign, gathering up in one all the remaining following after it, and Scorpio does the same

for the last four signs of the Zodiac, containing within
itself potentially all that is subsequently unfolded in
them.

We can realize that cosmically, at least, the sign Cancer
represents the mundane egg, so to speak, and thus contains all the seven-fold universe, synthetically, within it,
and on the other hand, the sign Scorpio may be compared
to an egg within an egg, and synthesizing the four lower
planes of the seven. These four are seen differentiated
in the signs Sagittarius, Capricorn, Aquarius and Pisces,
which relate by reflection to all lower quaternaries within
the various septenaries of Nature, and these are the planes
of form, as may be distinguished from the formless or
Nirvanic planes, represented by the signs Leo, Virgo and
Libra, and in the Orphic system they represent the sensible planes as distinguished from the intelligible or super-
sensible world of these same signs of Leo, Virgo and
Libra; therefore, while the signs from Leo onward each
relate to one of the seven cosmic planes, the sign Scorpio
is omitted in this relationship. It really forms a Zodiacal
eighth cast out and Libra corresponding to the third
plane, that is, from above, while Sagittarius corresponds
to the fourth plane.

The sign Scorpio resembles the monad in being the
starting point and finally unifying all things below it.
The real nature and characteristic of this sign, taken as a
whole, is both destructive and creative. In the same way
that Scorpio represents the stage at which the four lower
cosmic planes are differentiated out, and likewise conception is that at which the personality begins to differentiate in the physical manifestation, and in just the same
way as Scorpio synthesizes or gathers together the four
lower planes, so likewise, death unifies the operations of
the manifested personality in order that their fruition
may be gathered up in one into the soul.

We have here represented the cyclic procession of the

downward, the outward analytic arc; also the upward, inward, synthetic arc of the life circle.

The number eight or ogdad, symbolizes the eternal and spiral motion of cycles, its lines crossing in the center, running from above, to below, and back again to the above unity, and is thus symbolized in its turn by the Caducius. It represents the regular breathing of the Kosmos presided over by the seven great gods, composed as it is of two circles without beginning or end, represents the transference from one plane to another, that is, cyclic motion outwards and its return. It is also seven and one, or the octave; and aMrs, which rules the first sign Aries also rules its octave expression, that is, Scorpio; and to continue, Aries represents fire or the upward pointing triangle, and reflects downward, as Scorpio, watery sign, is the downward pointed triangle and water stands for matter in general, the substance out of which forms are created, and is also fruitful and productive.

This number eight expresses the equality of the human race and out of eternal unity and the mystic number seven, out of the heaven and the seven planets, also the spheres of the fixed stars in the philosophy of mathematics was born the ogdad.

In the eastern philosophy this number eight symbolizes equality of units, order, and symmetry in heaven, that is, the higher spheres, and transformed into inequality and confusion on earth, that is, the lower spheres, by selfishness who opposes Nature's decrees; but when perfect as a cosmic number it is also the symbol of the lower self, the animal nature in man.

The Figure Eight Symbolical of Eternity.

The figure eight is in reality symbolical of eternity itself for this reason, that is, the number seven represents every imaginable duration, while the unity which begins anew after the seven great planes of life's manifold conditions have been traversed, is beyond even duration. This figure eight represents the life eternal beyond all else—the life eternal which maintains itself by the equilibrium of motion, and this is all very aptly illustrated in this figure eight, forming, as it does, two juxtaposed circles. The figure six in form is found to be very closely associated with this figure eight.

The figure is composed of the half circle above with the circle below, eight of two circles without beginning or end, and representative of the equilibrium of life and being, of the idea and form, of the thought and word, of the light and shadow and also of spirit and matter. Both six and eight are even numbers, and are therefore feminine in this nature, and both imply duality, though there is a difference for the two triads belong to the formless, intelligible world, that is the two threes of which the figure six is formed, while the two tetrads or fours belong to the sensible world of form, and we may go on still further and study the close relationship that exists between the sixth and eighth signs of the Zodiac.

It is a peculiar fact that the second quaternary of signs, that is, Leo, Virgo, Libra and Scorpio are all, with the one exception of Libra, formed from two numbers only, mathematically. The sign Leo is the fifth sign, five is two plus three; Virgo the sixth sign is two multiplied by three; and Scorpio, the eighth sign, is two cubed. In the first, we have the monad, duad, triad and tetrad coming into expression; then the duad and triad as the mother and father combined to form three out of the next four numbers in various ways; and finally the triad and tetrad interpose number seven, though we must keep in mind that the tetrad or four is nothing more than the duad differentiated, and in this wise, it will be found that all the numbers come out of the duad and the triad.

The significations of the Virgo-Scorpio influence as embodied in one, in reference to human evolution, we have considered heretofore, and taking these two signs together as one represents the condition of humanity before the evolution of Manas or mind, the self-conscious intellect and the individuality, and their separation into two by the interposition of the sign Libra indicates the evolution of the Manas which separates, and yet at the same time unites the spiritual upon the planes above, to the animal man on the planes below, and then again, in a lower form of application, this influence also symbolizes the evolution of the sex from the asexual, passing through the androgynous state of being to the bi-sexual, a change which is said to have taken place in mankind prior to the evolution of the Manas. This sign Scorpio is said to signify the universe in thought, or rather the universe in the divine conception. This may seem strange but it is in actual accordance with what we have found in all the signs.

The figure eight represents much in its association with this science of life; and the sign Scorpio, the eighth sign of the Zodiac, is symbolical of both separation and unity as we have learned, and thus its importance is most significant, and in the perfected figure this sign is found at the apex of all others, and besides this, it holds the key to the mysteries beyond that mortal man has not penetrated. As we remarked heretofore, the sign Scorpio signifies the whole universe in the divine conception, and this may be better understood if we take the sign Leo as the highest cosmic aspect of the creative forces, and the signs Virgo Scorpio as primordial substance, it will be observed that Scorpio is the fifth sign from the ascendant and represents the fifth house, that is, in following the signs in order to the mid-heaven on toward the descendant. Then the interposition of Libra stands for the vibratory impulse of the creative force, which starts the primordial substance to differentiate.

We have learned that the great plan of the universe

existed in the divine mind long before it was manifested
in the material form, and the sign Scorpio represents the
potential stage, and thus the impulse is communicated,
and the four remaining signs are differentiated out and
signify the four planes upon which the universe in its
material form exists, the four planes upon which the
great evolution of man takes place, and when man has
passed these four planes he has attained the divine consciousness. Thus it will be observed that the separation
of the sixth sign from the eighth by the interposition of
Libra, that is, by its active manifestation, really symbolizes something much higher than the development of
the Manas in mankind. It indicates really the separation
and unity, the birth and career of Saturn as expressed
in the classical myth, and also signifies the active manifestations of those divine beings under whose fostering
care the animal man developed mind, which is signified
by the sign Capricorn.

The sign Libra, it is said esoterically, represents the
neoric order of the Gods in the Orphic theogony, and
this consisted of seven members, each one the head of a
septenary of powers. The seventh member was termed
Oceanus, and the separative Deity, that is, Scorpio, who
remained behind with his father Uranus.

When Saturn, whose exaltation is found in Libra, re-
belled and seized the kingdom and Saturn himself was
afterwards cast down into Capricorn by his own son
Jupiter in Sagittarius, and this casting down refers to the
fall of the angel Lucifer or Satan and the evolution of
the intellect in man.

Capricorn stands for the mental plane, the second from
above, of the four planes heretofore spoken of, and this
seventh monad is termed the separative Deity and is associated with the secret of Satan.

Then apart from this reference to human evolution, it
does not seem evident why a closer association should
be posited between the sixth and the eighth signs of the

Zodiac than between any other two signs that are similarly placed in the Zodiac. In a close observation, both numerically and astrologically, the signs Cancer and Taurus, the second and fourth signs, seem to be much more closely affiliated though there is no similar symbolism attached to them. In many ways these two signs are almost identical, but they are on the lower planes of the manifested life, though cosmically their separation by Gemini symbolizes in a way the descent of the Logos himself into matter, at the dawn of the universe, which is an architype of the subsequent fall of the promethean creators of humanity.

In our efforts to elucidate this divine science to humanity, we wish to do so in a simple form, and using terms that will be readily understood by the individual in his efforts to gather knowledge. In the consideration of figures it matters not whether we begin at unity and proceed toward the more complex, or this may be reversed and we may begin at the simpler numbers and advance towards the higher. With the first we may compare the process by which complexity arises out of simplicity with the evolution of a universe and its evolvement from homogeniety to heterogeniety, but when considering the method which shows the more complex series and reverting back to unity it corresponds to the death of the physical man or planet, star or universe and the resolution of the heterogeneous back again into primal simplicity. In this idea there underlies the belief in the efficacy or power and influence of numbers. This is most ancient in its origin as those individuals who dwelt in the golden age handed down to their posterity, but these truths, like many others, became confused with other material conceptions of divine wisdom, and finally became lost entirely to mortal minds.

CHAPTER X

The Number Nine Brings the First Cycle of Numbers to an End, or More Accurately Speaking, is the Introducer of the End.

We have heretofore considered the numbers seven and eight that are in a way separate from the others, and thus we have number six, the hexad, and pass on to the ennead, or number nine.

It is evident in the investigation of the principles of any whole, that when it is divided and sub-divided into a definite number of parts, each part will naturally reflect the whole within itself, and in a similar way each planet really contains all the elements within itself of the other planets; at the same time, its own proper characteristics and forces predominate, just as we have learned that each of the Zodiacal signs contains each and all the others reflected in its unity, and this same law is manifest in the expression of numbers, as for instance, if the more simple numbers are regarded as the true source and cause of the more complex, it is then quite evident that there can be nothing in the cause that is not really present in the effect and vice versa, for whatever is the effect must necessarily have been present in the cause, and this is to be found quite clearly demonstrated in the duad, or number two, as the duad, which is a dual unit, reflects the whole duality within each of its parts and immediately we may discern gives birth to number four.

Thus the numbers two and four bear a closer relation to each other than either the duad or tetrad does to the triad or number three, and it is in a similar manner that the two signs, Taurus and Cancer, are so much more closely in touch with one another than either is with the third sign, Gemini.

The numbers three and nine illustrate this principle quite aptly, as we may say three is a triadic unity, and when each one of its parts reflects the whole within it, the number nine, or ennead, at once arises and the triad gives birth to the ennead, as we may very aptly compare the two signs, Gemini, the third sign, with Sagittarius, the ninth sign.

The higher esoteric principle involved in this triadic unity is aptly illustrated by considering a large triangle containing three smaller ones, and if the small triangles are equal to each other and are arranged at equal distances, it will be seen that nine small triangles form one large one. Thus we may see, to begin with, that the number nine is derived directly from the triad, or number three, and really contains only in a more differentiated form, all that is present in its parent, the triad.

It may be well here to consider the close similarity and association between matters signified by the third and ninth houses in astrology; as for instance, we find the third house signifies short journeys, the ninth house long journeys; the third house letter writings and papers, the ninth house, books; the third house signifies ephemeral literature, the ninth house permanent literature; the third house signifies railways, roads, ferries, rivers, small lakes; the ninth house oceans, shipping, large bodies of water; the third house signifies intellect; the ninth house signifies spiritual intuition; the third house signifies the personality; the ninth house signifies the individuality; the third house signifies the brothers and sisters of the native; the ninth house signifies the brothers and sisters of the marriage partner; and so on through the various significations. As the triad, or number three, contains a manifested duality, we find Gemini is the first dual sign and the sign Sagittarius, which is also a double sign, we find to be the second triad and this duality may be described as the positive and the negative, the third point in the triangle partaking of the nature of both of these.

In consideration of the number nine, we have seen that in the duad we have both the positive and negative forces in manifestation, while the third point in the triangle represented by the triad partakes of the nature of both of these, and therefore in any triangle the apex may be represented as neutral, while the base itself differentiates and becomes triple at each extremity, the whole constituting a septenary which does not really become ninefold until the apex is also made triple, and in accordance with this it will be observed that under the old enumeration of the signs by which the three signs, Virgo, Libra and Scorpio, are regarded as one, the sign Sagittarius was then counted as the seventh sign of the Zodiac, while Capricorn was the eighth sign and Aquarius the ninth sign and there is in a way a manifestation of the natures of those houses expressed when these signs are placed in this wise, though the conditions of humanity in its evolvement has changed somewhat from the time the signs were used in the said manner, although even then all the qualities were present in the signs as they are today. For instance, if the sign Capricorn be considered as the eighth sign, we have the planet Mars the ruler of the sign Scorpio exalted in the sign Capricorn, the synthetic eighth sign. Then, again, we find a reason why the Hindus consider the sixth house to represent enemies of the native, as when the three signs, Virgo, Libra and Scorpio, are considered one then their powers would naturally be held in common, and in this way a natural characteristic of the differentiated seventh house might be attributed to the sixth house, etc.

Then, too, if the number nine, or ennead, is nothing more than the triad differentiating itself, the student would naturally expect to discover an even closer community of nature between the two than is held to be the case generally, and if the planet Mercury is strong in the first triangle it must naturally have much sympathy with the sign Sagittarius, which sign is directly obtained from this triangle, though modern students conclude that Mercury is weak and in its detriment when found in this sign, but in former investigations of the planet Mercury we find that Mercury is controvertible in nature and really takes upon itself the environments of the sign and house in which it is found; therefore this planet can never be out of touch with any of the signs for the reason that if it were it could never receive or respond to the influence of the planet ruling that sign, whatever it may be, and it will be found that Mercury in the sign Sagittarius will manifest a far different influence when in the ninth house in the radical position of Sagittarius than if found in the sixth, eighth or twelfth houses.

Then, again, if the sign Aquarius be considered as the ninth sign, we find it in close affinity with the sign Gemini both by triplicity and also sympathy of their rulers, as we have learned in considering the relationship existing between Aquarius and the sixth sign Virgo.

The number nine really posseses two principal characteristics, first it brings the first cycle of numbers to an end, and the various numbers starting from one and proceeding to number nine all possess peculiarities of their own when added or multiplied as we have seen in consideration of the table of numbers as given under the number eight.

As we are now dealing with the number nine, or ennead, and in further consideration of the chief characteristics of this number, we have learned how this number brings the first cycle to an end, though after the number nine has been passed, then taking the number ten by addition and multiplication simply reproduces one.

Thus the number nine becomes the end of the cycle, or more accurately speaking, the introducer of the end, as we find the real termination in the number ten or

decade of the first cycle, and in ancient times was termed
Telesphorus, interpreted, means a finality or bringing to
a close; just as we find the influence of a planet in its
progressive movement as it approaches the thirtieth degree of the sign through which it has been moving, it
already begins to take on the environments of the degree following, and the influence of the first degrees of
following signs are felt before the planet has really entered the sign. This is especially true of the swift-moving
planets, the Moon and Mercury; and the judgment must
ever take this fact into consideration.

We find the number nine is considered favorable, as
for instance, in case of sickness the ninth day the Moon
trines its own place at the beginning of the illness, and
is therefore casting a beneficient ray that tends to offset
the ill effects of the seventh day when the Moon squares
its own place.

Then, again, the number nine gives perfection and
consummation to the offspring that are produced in nine
months, referring, of course, to the human family who
have developed to a consciousness of being and reasoning; at least the child is ready to take up the link in the
chain.

The number nine is odd, corresponding to the male;
the number eight is even, corresponding to the female,
and this is significant in the manifestation of life of sex.

The ennead is also considered the sacred number of
being, and becoming the triple triad or nine, is followed
by the monad, or number one, making ten. Thus the
unit, or monad, is really that which is and becomes, and
the triple triad is that which enables it to be and become;
therefore nine is the producer and ten the produced; in
just the same manner as Cancer, the great mother sign,
is the ninth sign from Scorpio, while Leo, the child sign,
then becomes the tenth.

Then, again, we may consider the ninth sign, Sagittarius, the spiritual channel for the evolution of the hu-

man ego, and thus it corresponds to the ennead in numbers and is followed by the sign Capricorn, representing the ego itself, and is the tenth sign terminating this first cycle in its experience.

If we consider the number six as the symbol of this Earth planet ready to be animated by a divine spirit, then the number nine symbolized this globe ready to be informed by an evil spirit or influence, that is to say, the three triangles are not complete without the unit which follows after and completes them, thus forming the decade, and therefore the number nine implies a tendency to be dragged downwards into number ten just in a similar way as the note corresponding to B, which expresses a restlessness, corresponding to Mercury, that requires the key note C or do to follow in order to complete it.

The Number Nine is the Great Symbol of Materialization.

Then, again, three multiplied by three is the great symbol of corporization or the materialization of spirit as was taught by Pythagoras, meaning a bringing down from the higher planes of spirit into gross matter, though it is applicable as a general principle to other planes than the material plane, and may be applied really to the first seven planes of life.

The qualities of the will are being expressed in the lower planes of life that is so important to the growth, development and realization of the human family.

So far as its effects are concerned, it may be compared to the ninth sign of the Zodiac, Sagittarius, the centaur that shoots his arrows with unerring accuracy, and if we would express the idea in the abstract form, we may consider that in any cycle the last unit but one carries with it the tendency to complete the cycle, and this tendency may be upwards or downwards, good or bad, spiritual or material, just according to the quality of the cycle and its mode of application.

This, then, is the arrow of the centaur, the irresistible impulse onwards towards the goal to which each soul is consciously or unconsciously to the lower mind seeking to attain, and thus for the same reason the number nine, or ennead, represents a symbol of great and important changes, cosmic and social in nature and covering a wide field of subjects.

It is also what was termed the sad emblem of the fragility of the human race, for the reason that it is incomplete until it has passed onward into the decade, and it also is symbolical of the act of reproduction and generation, as there is found to be the same completion of a cycle by the introduction of another unit, and at the same time it is evident why the ancients considered all circular lines to be represented by this number nine, thus being sacred to spheres.

It stands as the sign of circumference, since its value in degrees is equal to the number nine, or as we may say in sub-dividing three, plus six, plus naught.

Hence, it is also the symbol of the human head, and it will be observed that the number nine is formed from a circle and a half, showing that the impulse is onward but is not yet complete, as we have seen in the number eight of the two circles joined, but even this does not express all that is to be expressed to the individual, and it is necessary to have the monad, or one, represented by the straight line and the naught or circle both separate ere it can be truly expressed.

The ancients termed the number nine the ocean, for the reason that it contained so much and was so difficult to fathom or exhaust all that was to be found therein, and Pythagoras speaks of it as flowing around the other numbers within the decade similar to the ocean, and this coincides also with some of the interpretations of the ninth house, voyages, shipping, etc.

This number nine was also termed the horizon, as the ocean was pictured as the boundary of the solid Earth, surrounding it on all sides just as the horizon bounds the observer's view in every direction.

Then another term that was applied to this number was Prometheus, the same that was considered to have stolen fire from heaven and gave it to mortals, and for this reason the number nine may be said to steal or absorb all the characteristics of the numbers that precede it and to hand them onward to those that follow it, for the second cycle of numbers simply reproduces the characteristics of the first, which was brought to a termination by the number nine; and for this reason it was compared to Terpsichore, the muse of dancing, for the reason that one cycle follows after another similar to the rythmic movements of a dance.

The ninth sign, Sagittarius, signifies religion, prophecy, genius, intuition, spiritual knowledge and foresight.

The nine muses synthesized in Apollo, the tenth were all types of genius, as considered by the ancients.

The nine muses were considered to have been the daughters of Jupiter, and this is significant as Jupiter has rule over the ninth sign of the Zodiac, Sagittarius and Mnemosyne, or rather Memory was the mother. Thus memory, being reminiscence or instructive, memory manifesting in various expressions of the individual.

Each of these muses corresponded to a distinct number, one of the numerals, as for instance, Terpsichore corresponds to the sphere of Jupiter on one side, and to the number nine on the other, while both Jupiter and the number nine belong to the sign Sagittarius. Apollo, the leader of the nine, corresponds to the monad. The word itself interpreted means not many, hence a unit. Apollo is the Sun god, and we find the Sun is exalted in the sign Aries, the first sign of the Zodiac.

Erato, the muse of amatory poetry, corresponds to the planet Venus and the number two, and these both manifest in the sign Taurus, the second sign of the Zodiac. Erato was invoked by the Romans in the latter portion of the month of April when the Sun enters the sign Taurus.

Melpomene, the muse of tragedy, corresponds to number five and the Sun, and these both find their manifestation through the sign Leo, the fifth sign of the Zodiac, and so on.

It will be observed that what the number nine is among numbers, that the planet Jupiter is in mythology. He is the Demi-urgus, or the fabricator of the sensible world of form and in exactly the same manner as Sagittarius is the highest sign of the four that correspond to the four lower cosmic planes, so Jupiter is the lord of the ninefold super cosmic order in Orphic theogony, and Jupiter is said to swallow Phanes, and this is the ideal plan of the universe as it exists in the mind of the Logos is taken by the hierarchy of builders, or fabricators, as the model upon which they are to work; and by a careful investigation this same character will be found to be expressed through the sign Sagittarius, as for instance, when this sign is represented in numbers, the name is equivalent to the number nine, and the division in question, counting of course from the first sign, Aries, or Musha, while the sign Sagittarius is termed Dhanus, and this sign indicates quite clearly the nine Brahmas, that is, the nine prapapatis who assisted the Demi-urgus in constructing the material universe.

The ennead, or number nine, as we have already observed, is characterized in just the same way, that is, it takes the powers of the earlier numbers as itself and then hands them onward to those that follow it, or as we may say in another number, it really constructs the second cycle of numbers after the model contained or manifested in the first cycle except that it lifts them up to a higher expression and consequently to a higher plane of life.

We have seen that Saturn is the father of Jupiter, though Jupiter is spoken of as the father of Gods and men and reigns as supreme in Olympus. This is so for the reason that none of the powers anterior to Jupiter can be effective in the manifested world, except through the hierarchy represented by him. Thus he is supreme in that world and really gives birth therein just as Buddhi in man, to which the sign Sagittarius corresponds, is the vehicle for atomic life upon the lower planes.

In the system of Orpheus the super-cosmic order of gods were composed of three triads, with the planet Jupiter as the head or governing power. They were first the Demi-urgic triad of celestial Jupiter, Neptune or the water Jupiter, and Pluto or subterrian Jupiter.

The second triad was termed the Carybantic triad of Diana, Proserpine and Minerva; and the third triad referred to, the triple Apollo Proserpine, means care, and the Carybantic triad is analogous to the Curetic triad in the Saturnine order. The two names, Proserpine and Curetis, were applied directly to the number nine by the Pythagoreans, as this number possessed these appellations in consequence of its consisting of three triads, the triad harmonizing both with the Curetis and Prosperine.

This number nine also represents what the Catholic theologians term the circumincession of the divine persons. *Circuminsessio,* interpreted, means the power of residing around each other without confusion of the conceptions. In the number nine we may obtain a more clear comprehension of what is meant by the number three differentiating itself whereby each of its parts reflects the whole contained within it, thus forming the triple triad, the three triangles in one.

This number nine may also be said to be the number

of initiation, and this is in exact correspondence with the number nine and the Zodiacal sign Sagittarius to the spiritual or Buddhic plane, that is, the highest of the lower four contained in the cosmos.

The Zodiacal tenth sign Capricorn represents the principle of individuality in man, that is, the self-centred human ego, which evolves from the highest principle downward into matter and finds expression through all the various gradations of matter up to a consciousness of being, and this self-centred human ego finds its spiritual home or affinity on the mental or Devachanic plane, to which we have heretofore referred. and this corresponds to the sign Sagittarius, and this sign is thus one stage higher, the spiritual plane and residing place of the principle termed Buddhi in man, and the signs Leo, Virgo and Libra that follow the egg as contained in the sign Cancer representing the atma, while the mode of consciousness corresponding to the sign Capricorn is that of intellectual self-consciousness, that is, the limited and separate self; but on the Buddhic plane it takes the form of the blending of these separated selves in one, at the same time these selves do not lose their individuality, which may seem somewhat contradictory, but nevertheless it is a reality and marks one stage in the evolvement of the human soul.

This is a mode of consciousness corresponding to the planet Jupiter and the sign Sagittarius, and here again we may realize why this sign, the ninth, in its radical position might be numbered as the seventh in the synthetic-Zodiacal and also how the general nature and characteristics of the seventh house should be in some respects applicable to the sign Sagittarius, as it represents a state of being which each is himself, and with it a realization that cannot be found upon the lower planes of life, and at the same time each individual feels that he is one with the whole inseparate and inseparable, and the nearest comparison to this condition on the lower physical plane is the condition that exists between two persons who are united by a pure, intense love, which makes them feel as one and causes them to live, act, feel and live as one, recognizing no barrier, no variance, no separation.

We have learned of the plane of life into which the individual passes after his first true spiritual initiation, a plane where each is separate, though all are one in thought, and comprehend thoughts and ideas with the same interpretation, and it is the influence that is sent out from this plane which causes mortals to seek happiness by union between themselves, and also the object of their desire, no matter what that object or desire may be. By a comparison of these conditions we find this corresponds to the signs of Capricorn and Sagittarius, in that the consciousness is lifted up out of Capricorn into the sign Sagittarius, and also out of Saturn into Jupiter.

We have observed that the sign Sagittarius is a double sign, and it is also a fact that this mode of consciousness is a manifestation of the two in one or indicating the second person of the divine trinity or Ananda, that is, a condition of love, bliss and harmony. This plane of life may be termed Fohatic. The word Fohat interpreted signifies the active or male potency of the shakti, that is, the female reproductive power in Nature. It is, as we may say, the creative cosmic impulse differentiated into polarized positive and negative forces and directed by the intelligences back of it.

Then again we may find it expressed as the six forces of Nature in the sign Virgo, the radical sixth sign, at least potentially, though it is rendered active in the hierarchy of builders represented by the sign Sagittarius. We find then that the number two is only potentially dual in its manifestation on the lower planes of life, while in the number three or triad we find active duality is expressed. Then we may say that what the two triangles of Virgo number six contain potentially, the three

triads contained in the sign Sagittarius manifest actively.
Then again the term Fohat may be described as the essence of cosmic electricity, and in association with this
we may look upon Jupiter as the God of lightning and
the thunder bolt, and in connection with this, there is to
be found in the mystical hymns of Orpheus, one to thun-
dering Jupiter, and another to Jupiter as the primary
cause of lightning, and herein we may find a connection
between Jupiter as the ruler of Sagittarius, also the ruler
or part ruler of the sign Pisces, one belonging to the
fiery triplicity, and one to the watery triplicity, the ninth
and twelfth signs both representative of the triad and
both active; though one is outwardly active while Pisces
is active from within, but one leads to the manifestation
of the other, and while it is possible to comprehend the
higher interpretations of the sign Sagittarius it is difficult to realize the higher meaning of the sign Pisces, at
least to those on the lower planes of life.

The expression Fohat is the steed, and thought is the
rider, is also significant of the sign Sagittarius, and Fohat is also spoken of as the fiery serpent which hisses as
it glides, and teth, the ninth letter of the Hebrew alpha-
bet is a serpent. Then, too, Fohat is connected with
Eros the divine desire of Leo with the primordial impulse of Aries, and with the suddenly acting electric
force of Urania and Aquarius, thus representing the positive or vital side of spirit matter. It has a bearing upon
all the odd signs. Four and nine are the only two square
numbers within the limits of the twelve, and Sagittarius
represents the fourth plane within the egg of Cancer and
Jupiter ruler of Sagittarius is exalted in Cancer.

CHAPTER XI

Number Ten In-gathers, Synthesizes, Returns and De-stroys, to Rebuild Anew Upon a Higher Plane of Life.

We shall now take up the number to which we have referred heretofore as the decade. This, we must recognize, means a ten in one, and therefore will naturally contain and possess within itself as in a microcosm the powers and qualities of the whole cycle of numbers, and this fact may readily be observed in a consideration of the tenth sign of the Zodiac, that is, Capricorn, which sign corresponds to the decade or number ten.

It will be seen that the sign Capricorn stands for mankind, and for the evolvement of that individual consciousness of being that naturally distinguishes man from the animal, for as we have stated heretofore, man is a microcosm in himself, and really possesses within himself the potentiality of growing into the likeness and wisdom of the Creator, as it has been claimed that man was created in the image of the Gods.

We may observe that the number one starts the cycle and the number ten terminates the cycle, and the two numbers merging into one at that imaginary line of division between them, where there is neither beginning or end; in other words, that point where the monad, and the decade, are the same under different aspects, and it is from this point that number one sends forth, differentiates, evolves, creates, and to this same point the number ten in-gathers, synthesizes, returns and destroys to rebuild anew upon a higher plane of life, which is to be accomplished from the cycle that is to follow.

In the Zodiacal signs, the first sign, Aries, corresponding to number one, is found to be creative, hot, expan-
sive in nature; while the tenth sign, Capricorn, is found
to be cold, contracting, and in a sense limited and destructive in qualities.

Then, too, it will be observed that the planet Mars
which is ruler of the first sign Aries, is exalted in Capricorn the tenth sign. This is significant, and an arithmetical consideration of the number ten demonstrates
that it is manifesting many of the principles as expressed
in number one, for in reality they are only outwardly
varying aspects of an underlying unity. It naturally follows from this that any cycle may be interpreted and analyzed on a scale of ten quite irrespective of the number
of parts apparently contained within it, for the cycle in
terms of numbers is not really complete until ten is
reached, and to proceed further in their enumeration, is
to begin anew.

We find the circle of the Zodiac, the horoscopal figure,
is divided on a scale of twelve, first into quarters which
possess their own peculiar significance, and then subdivided into twelve equal parts, though it will be found
possible to use the scale of ten, or indeed any other number proceeding by an equal division from the first degree
of Aries, or the degree upon the cusp of the ascendant,
whatever that may be, and the segments of the arc thus
obtained, would then be found to possess qualities and
powers that would be determined by the scale of division
employed, and the same rule holds true of the division
of the signs of the Zodiac. A Zodiacal may apparently
be divided in a great variety of ways, each method having its own use, application and interpretation.

Some of these are found in the ancient teachings, some
of which have been handed down to the Hindus, but outside of this are scarcely referred to, as they have been
lost through the process of time, and perversion.

We may observe that the sign Capricorn is associated

with the God Pan who was identified with Phanes by the Orphic school.

The number ten signifies one complete cycle and stands for a perfect whole. It denotes completion, and thus any one being or thing that is finished or is complete within itself, may be compared to the decade, and in its significance upon the higher spheres, it may be termed the cosmos, tenfold and symmetrical, and on the lower planes it refers to man, also tenfold.

As we have shown heretofore, Phanes is associated with the Pentad and fifth sign Leo, that is, in its higher interpretation of the cosmic and spirit life, and the God Pan is associated with the decade, and the sign Capricorn, the mortal application of the same idea, that is, a microcopic Phanes. As man, the microcosm is the mirror of the universe, the macrocosm.

We find the decade is formed of two pentads or two fives in the same manner as man is a pentad within a pentad. We also find that Capricorn, the radical tenth sign, is the fifth sign within the egg or sign Cancer, and according to the table as given heretofore under the number six, we find Saturn the ruler of Capricorn belongs to the fifth evolutionary chain in the solar system, that is, counting from Vulcan outwards, and the resemblance is quite marked between the figure five and the symbol of Saturn, or what has been termed the sickle of Saturn, and in reality the figure five owes its construction to the genetrix of the earliest ages, the mother goddess of time itself, and this figure five is a form of the crooked sickle of Kronos that was derived from the Khepsh thigh of the hippopotamus, which animal represented the constellation of the great bear.

This is, then, quite significant, as it shows this number five to be a sign of cutting off, and in its physiological sense, this cutting off is that of the child from the mother, and a direct reference to the fifth mundane house, but when considered in a broader sense it implies a separation, a making definite, a completion, as Phanes completes and synthesizes the cosmos, while Manas the fifth principle does the same in man, and Makarain, the Hindu term for Capricorn, interpreted means five sided Ma, being five in Sanskrit from which the word is derived, and we may observe that the dodecahedron is bounded by pentagonal faces and the signification of the term Makaram is the shape of the material universe in the mind of the demiurgus, that is the dodecahedron, and the human significance of the five pointed star is well known, its reference being outwardly to the head, arms and limbs of man and inwardly to Manas itself.

Pan and Phanes are one, the all father, the Logos, and the cosmic application of the pentad or number five is found in the sign Leo and the human in Capricorn.

In the sign Leo we have the cosmic tetrakys giving birth to the universe, the child, while in the sign Capricorn, Atma with its three hypostases and Budhi give birth to Manas the human fifth principle, and this principle limits, completes, and makes definite the human mind, or rather the consciousness, and before the advent of man consciousness in the animal, vegetable, mineral and elemental kingdoms was monadic only, that is, it was unindividualized, but with the evolution of the human ego, self consciousness is manifested, individualized, limited and self-centered. The sign Capricorn expresses this evolution and limitation and Saturn its ruler thus stands as a type of man.

We have seen how the sign Capricorn, corresponding to the number ten, expresses the evolution and limitation though it is essential to have the cold rays of its ruler Saturn manifesting through it, as the soil, in order to realize more clearly the operation of these forces in conjunction with their effects upon all matter forms, etc.

The number ten representing the complete circle, stands for any cycle whatever it may be, and therefore for time in general and Chronos.

Saturn is known as the god of time, which includes all cycles, and the correspondence here is quite evident for the cusp of the tenth house which is termed the mid-heaven of the horoscope, is in fact the mundane representative of both Saturn and the sign Capricorn, and this is used generally as the measure of time in astrological judgments and interpretations, while the first house, or ascendant, where the signs rise irregularly, marks the character, individuality, etc., and the mid-heaven, to which they pass quite regularly, marks off time.

Then, in a sense, too, Manas which has its habitat on the upper mental plane, performs the same office in the evolution of the human race, for it is to this condition that the soul finds expression after so-called death of the physical, each out-breath and in-drawn personality marking off a small and subsidiary cycle of time in the long career of the ego itself.

In the Egyptian planisphere the sybmol of Capricorn was not that of the goat; instead it was an animal with the head and two front legs of an antelope, while the body and tail was that of a fish, and we find this sign Capricorn also associated with the Egyptian and Hindu crocodile symbology. We learn that Sevekh, the seventh, was the crocodile type and he attained the dignity of the first god as Swekh Kronus, the dragon, and in the African or Kaffir dialects the crocodile and the soul are synonymous as they were in Swekh on account of his superior intelligence.

It may be observed that the soul of Seb the schthyphallic father of the fifth creation is identical with the Buddhist fifth principle, which principle is Manas, the human soul, and in the correspondence of planets and principles we may consider Saturn as the ruler of the lower Manas and Urania of the higher, and Saturn as we know is the ruler of Capricorn, while Urania is the ruler of Aquarius, over which sign Saturn is also considered to have some influence in his higher manifestation.

We may also observe that in the consideration of the ancient seven planets when the Sun and Moon are both included, the planet Saturn is the seventh, but when these are omitted then Saturn becomes the fifth as we have already noted in its relation to the pentad, and Seb, whose name signifies the number five was the lord of the fifth creation. He is a star god and still was termed by the ancients the god of Earth. He is the Egyptian Saturn and the goose was sacred to him, and he was considered to have laid the egg of the world.

This in its human and higher application evolves into the body of Manas, that is, the casual body, and this refers to Cancer as the cosmic egg, and Capricorn as the human, while the one encloses and shuts in, defines and limits the universe and the other does the same thing for man, and it is significant that we find Cancer, the mundane fourth, in opposition to Capricorn the mundane tenth house in the horoscopal figure.

We have already learned of the limiting, contracting and defining influence of the ruler of this sign Capricorn, and this very process constitutes the distinction between monadic consciousness as manifested in the animal and vegetable kingdoms, and the specialized self-consciousness of man. Thus we may observe that the consciousness that is general and diffused has to be enclosed or confined within certain limits, in order to mark off self from not self and thus build up the individuality.

This is a law of necessity in the evolutionary process as may be realized by a close study of these conditions, and this may be readily recognized as the influence of the planet Saturn and the very kind of evolution in terms of consciousness, that truly belong to the Zodiacal sign Capricorn.

In a similar way as the two gods Pan and Phanes are identified, both representing the All Father, the Supreme

Intelligence, while at the same time each one has its own human correspondence, and so the crocodile in its cosmic reference signifies the Sun, corresponding to the heart and the brain of the macrocosm, while in its human application it stands for the soul, and the significance of the crocodile coming up out of the water signifies the Sun rising out of the great deep of space on the one hand and is also significant of the human soul arising out of the turbid waters of terrestrial life. This arising or resurrection takes place at the so-called death of the physical, and this is termed the iniation into the higher or terrestrial life, and from this the sign Capricorn with Saturn and the crocodile and the fish myths of ancient times were intended to denote the immortality of the soul, and its triumph over the mere physical existence, a triumph that is lasting and permanent in its effect; and referring to the myths of the fish of which Jonah's is a type were really expressions of this idea of iniation.

It may be a little difficult for the student to realize the relationship of the fish of Capricorn to that of the sign Pisces, though it will be observed that when the three signs Virgo, Libra and Scorpio are merged into one, this would bring the sign Pisces as the tenth sign and Capricorn as the eighth and there was a time in this Earth's evolution when this classification would better express the forces then existing and contained in the signs of the Zodias, and it is significant that the last five Zodiacal signs are more or less associated with water, as Scorpio is a watery sign.

Sagittarius belongs to number nine, termed the ocean, and the significations of the ninth house which has to do with long voyages and shipping. Capricorn was represented as half fish and Aquarius in its symbology shows the water bearer, while Pisces is one of the water triplicity. Water is a condition of matter, in a higher expression. Manas is dual and the latter part of Capricorn rep-

resented by the fish corresponds to the lower Manas, that
which descends into physical life, and which is signified
by the sign Pisces; and the first half, the fore part of
the antelope, corresponding to the higher Manas; and
the goat-man Pan, and the fish-man whose lower half is
a fish, or who is depicted emerging from the mouth of a
fish, or wearing a fish skin as a robe, are all to be given
the same interpretation.

This number ten is composed of the luminous unity
and the dark Zero and there are two pentagrams in this
number as there are two triangles in the number six.

The two pentagrams contained in the decade are varied in their vibration, one five being deeply colored, the
other five being transparent or crystal like; or in other
words, five are impure and five are pure in their expression, five races of giants and five angels who oppose
them. Then we have learned of the five foolish virgins
and the five wise ones, etc.

It may also have been observed the close resemblance
that exists between the sign Leo as the first fifth sign
and the sign Capricorn as the second fifth sign, for both
are found to be signs of rulership, power, and gives
authority over others, while the rulers, the Sun and Saturn, stand for the macrocosm and the microcosm, that is,
the Sun for the universe and Saturn for man. The Sun
and the tenth house both signify the ruler of a country
in mundane astrology, and while the fifth house is known
to signify the offspring. In mundane the environments
of the fifth house, at least Leo, must be noted in order to
determine the complete judgment of power, etc.

Then again, their relationship is strongly signified in the
fact that if the first point of Aries, or the ascendant, and
its cusp be taken as the point of conception, the sign Capricorn, or the cusp of the tenth house, would represent
the physical manifestation of the child in the lunar time;
and the Chaldeans recognized the association of the

planet Saturn with the Sun, that is, they perceived the
necessity of his cold rays in order to limit the influence
that would develop too quickly certain qualities of consciousness that without such influence would be out of
balance.

CHAPTER XII

The Nuptial Number of Plato, Ten Multiplied by Five.

We may here consider the nuptial number of Plato which is fifty, that is, as considered by him, ten multiplied by five, though he arrived at the results in this wise: If the numbers three and four represent the upright and the base lines of a right angle triangle, the hypothenuse will be five, according to the equation; three squared, plus four squared equal five squared. This in complete form is nine plus sixteen equal twenty-five and the sum of these numbers is fifty. In this consideration, three is the father number, and four the mother number, and five is the child; and the numbers four and five here receive their proper characteristics according to their mundane houses, and under the consideration of the number five we have learned that the number three is often associated with Mercury, Hermes and the father, and it will be observed that an indefinite series of similar equations may be drawn out, of which it will be found that the father numbers are ever the odd numbers taken in the series, while the mother numbers are ever the so-termed triangular numbers multiplied by four, in their regular order.

Then we may observe that while the number ten brings the first cycle of numbers to an end, the Zodiacal signs are twelve fold, and in order to complete this part of the subject it would be essential to investigate the hendecad and the dodecad, their relation to the signs Aquarius the eleventh, and Pisces the twelfth sign.

However we may consider this subject in the relationship of numbers upon the Zodiac, in its astrological and mythological associations and interpretations, it is so vast that it is impossible to touch upon the details, though our purpose is to demonstrate that this relationship of these numbers, planets and signs of the Zodiac is a reality in the great infinite scheme of the universe, an important truth, and not a mere accident or coincidence, for everywhere in the universe we may observe method and system, law and order, and perfect arrangement of the whole that is beyond human comprehension which could not find existence except through the infinite intelligences who create, sustain and supervise the same. The solar system in its visible and invisible unity has been brought into being in accord with this plan.

This subject is of importance in obtaining an esoteric comprehension of the meaning of these numbers and their associations with the universe of spirit and matter and the intelligences which brought into manifestation the various forms.

This solar system in both its visible and invisible aspects, being brought into expression or existence in accordance with the plan that was pre-existent in the mind of its creators, and we may go back in thought to the dawn of such manifestations when form did not exist save on the one hand, the Logos, and on the other, that incomprehensible root of what afterwards was brought into manifestation as matter forms, and for instance, we may be able to realize a creative idea arising in the mind of the Logos, the great infinite plan of the universe as it was to be, complete in all its divine magnificence, self rounded and embracing all things, at the same time including all things that are to be in the universe from the first, corresponding to unity, the monad, to the end corresponding to the decade, and this creative idea is the model upon which the whole universe in all its immensity is builded.

At the same time it not only serves as a plan for the system as a whole, but is applied with the necessary modifications to every part and subdivision of it, from the highest vibration in the celestial regions, down to the most minute atom of which the physical globes are created and formed.

Just what that great plan is in its creative simplicity as it exists in the divine mind we cannot fully comprehend, for it is at the same time too simple, too abstract, aye, too vast for the finite mind to comprehend though it may be easily demonstrated that wherever the mind can search, analyze and classify, from the highest esoteric to the lowest exoteric and concrete subjects in Nature life and consciousness, there is to be found evidence of intelligent arrangement, and that all seems directed towards a common point or stage in the evolvement and in spirit by a common motive, and this creative idea, impressed upon the spiritual root of all matter, sets the universe evolving while the grand intelligences back of all are themselves brought into existence in accordance with this great plan of being, and then taking up in their turn the great task of superintending the various kingdoms of Nature that are allotted to them, they evolve all subordinate things according to this one primordial plan, and in accordance with this divine plan, the worlds are created and made to revolve round the central sun in their particular motion and orbit, while their inhabitants, human, super-human and sub-human follow the same unchangeable course in accordance with the law, not only in the mode of their being, but also in the manner of their evolution, and it is a truth that there is no department of being, animate or inanimate, there is no sphere or plane of human activity, motion or reason, there is no science or philosophy or even a religion that is outside of its all embracing scope, and in accordance with it, mankind comes into physical expression and finally passes on to a higher manifestation, to a plane better adapted to his further progress, and thus nations appear in their manifestation and then when their part in the great scheme is completed they disappear and others come to take up where they left off.

It is in accordance with this great divine plan that worlds are born and decay in time, as the great waves pass through and onward in their evolvement of the whole. This great divine plan moulds all things, imposes bounds and limits upon all beings in manifestation.

In the consideration of the great infinite scheme of the universe under the control and guidance of an all-wise Intelligence we must realize clearly that if no such comprehensive, all embracing plan existed, the uniformity of law in all regions of being would be impossible.

Mathematics the Only Exact Science, at the Same Time the Most Profound and Mystical.

However, when, after a careful study and investigation of these universal conditions, a uniformity has been found to exist in the known, the unknown may be classified in terms of the known, and although as we have said, the finite mind is too small to contain the whole of the divine plan in all its spiritual abstractness, on the other hand the more we examine and analyze this great phenomena, and the further we push our classification, the more confident will the student become of the actual existence of such an all-wise plan, and from it the student may draw analogies, comparisons, correspondences which sometimes appear fantastic when viewed from the exoteric side only, are in truth real, vital, and are actually based upon principles that can be found to exist in the nature law.

We may observe that mathematics is the only exact science, and still at the same time it is the most profound and mystical, more especially when taken from an esoteric point of view, and considered in the interpretations of the numbers as heretofore given; while to the finite mind, in its efforts to grasp these interpretations, there may appear to exist contradiction, still a careful study and esoteric analysis of the subject will demonstrate that it is not so, but is a reality manifesting in perfect harmony and accord, at the same time being perfectly accurate in their various expressions, and fulfilling their mission in the great universal scheme; for we may find oppositions in the concrete as these are essential to create activity in the concrete manifestation. As we realize in the study of the Mars and Saturnine vibrations, and an investigation of the polar opposition and duality to be found in the influences as manifested by the signs and planets possessing these peculiar qualities, it will be observed that the dual signs are odd and masculine in their nature. This is significant in the study of the character of the signs, in order to judge correctly of their peculiar influences upon the house in which they are placed in the horoscopal figure.

Then we may depend upon the accuracy of the influences and those who imagine that accuracy and definiteness are incompatible with these occult and mystical influences, are very much mistaken in their calculations of these forces, when they are so ready and willing to judge merely from the objective point of view and to measure with lines laid down by scientific research and have never studied the matter only from this side.

There are many truths contained in the universe and man which may be at the present time vaguely and symbolically understood in some instances, for the reason that they have not been wholly and completely discovered, and in others because they are so vast that there is no likelihood of their being completely comprehended by the finite mind.

Then again, the idea that the mathematical science has no further interpretation than the simple straightforward one which lies on the surface is erroneous. What can be more simple than the statement that one and one are two.

The mere multiplication or addition of one and one is very simple, but on the other hand when this same problem is translated into philosophical terms, the infinite dividing and becoming finite, the unlimited giving birth to the limited, the unconditioned becoming conditioned, what can be more profound and incomprehensible.

As mathematics is a distinct branch of human knowledge it is certainly right and proper that the mind should seek to understand what it can, of the divine plan in the various natures and characteristics of numbers and forms, and then follows naturally the task of relating this to what may be discovered in the same plan of the Zodiac, and also the planets, as well as in the Zodiacal interpretation of various mythologies.

CHAPTER XIII

The Constitution of the Solar System Analyzed in its Invisible Planes of Being.

The interpretation of the Zodiac is of itself not a final truth; it is rather an attempted correspondence between the plan of the Zodiac, and that of what little the finite mind can comprehend in nature.

The constitution of the solar system may be studied in invisible planes of being, and outwardly in the arrangement of the Sun and planets on the visible or physical planes. Then the fixed stars themselves, so far removed as they are from the earth planet, are necessarily included in the calculation, for they form, so to speak, the environment of this solar system in just the same way as an individual animal or plant may be studied from the point of view of the influence exercised upon them by their various surroundings. The solar system, when taken as a complete unit, may be regarded as conditioned by its stellar environments, even though the space lying between is incomprehensible to the finite mind; and in going back to a time in the evolution of the universe when our system had not begun its existnece we may imagine what would seem to be void space, the Logus of the future solar system, and crossing and re-crossing in all directions were lines of influence from millions of distant suns and stars, and conditioned by such environment in the same manner as the plant is by the climate. This new system grows into being and becomes manifest in matter form and its stellar surroundings are capable of being grouped and classified in a systematic and methodical manner according to the influence exerted upon it, as the four winds of heaven may be classified according to the favorable or unfavorable influences they may have upon a growing plant.

The constellations or groups of stars so arrived at, surround the Earth, aye this solar system on all sides, above, below, similar to a hollow sphere, and the real stellar Zodiac is therefore spherical in form. However for the reason that the planets revolve around the Sun in what may be termed one plane, a central band of constellations is thus marked out by that plane and gaining in this way an importance beyond the rest we may term it the Zodiacal constellations, not the Zodiacal signs, and in this stellar Zodiac the student may imagine the horoscope not of one individual, of one nation, or one planet, only to be written, but the horoscopal figure of the entire solar system with all its periods, cycles, crises and changes of the greatest cosmic importance in the evolution of the same.

In the influences as cast by the stellar forces upon the planes in spirit life, it is essential to become familiar with the nature and character of each one of these constellations referred to, that is the constellations of the spherical stellar Zodiac, and in order to comprehend the astrology that applies to the planes in spirit life, these will have to be accurately known and classified, those above, those below, so to speak, as well as the central band, and then not from the point of view of the modern astronomer only, useful as this may be, but from the consideration of the true occultist, who can realize why one star belongs to one constellation rather than to another, and why a constellation is outlined in one way only and not in one of a hundred other ways that might be suggested; and this task of mapping out the constellations is one that will require careful study and time to accomplish, and one that will require the combined efforts of the true occultist and scientist to accomplish.

However, it must be realized that the first important consideration to comprehend is the influence of the planets and signs that fall directly upon the physical form and object, and it is this with which the individual is most concerned. We may realize that the constellations of the spherical Zodiac are the true condensers or media of expression from the higher planetary influences, and these higher, finer influences then find expression upon the material plane through the physical vibrations of the planets as manifested through the Zodiacal signs.

Thus it may be realized that every physical manifestation is first manifested in the spiritual before it reaches the matter world, and in order to discover the influence that is cast upon the higher spiritual planes a study of these great constellations must be made, to which the astrology of the material world is most simple in comparison, and the student can realize just what an infinite science he is dealing with. We will not go into interpretations of any of these influences, the object being only to show the immensity of this vast science of life eternal; and then, too, there is another consideration that is of importance, that is, the allowing for the possible change in the nature of the constellation, owing to the proper motion of the stars among themselves, which change would affect this grouping after great cycles of time, and such change would have its effect upon the influences cast upon mundane astrology as well as all the physical branches of this divine science.

We have spoken before of the relationship of numbers with the Zodiacal signs, and in this way it may be utilized practically for prediction, as for instance, the monad or number one, may be regarded as possessing the powers of its Zodiacal first sign, Aries, and the duad, or number two possessing the power of its corresponding second sign of the Zodiac, Taurus, and the triad, or number three possessing the powers of its corresponding Zodiacal sign, Gemini, and we may continue on through the numbers and signs in the same way up to the number twelve, and its corresponding Zodiacal sign Pisces, and using these numbers as equivalent to the corresponding signs and their ruling planets it is possible to elaborate a predictive system that will be based upon Zodiacal cycles in which the sign or planet will indicate an event while its corresponding number will point out the year in which it will likely take place.

It is true that these Zodiacal cycles vary in each horoscope just according to the sign rising and occupying the confines of the first house together with the ruling planets and following in their interpretation the ordinary rules as given in the science of mundane or natal astrology, as for instance the number associated with the sign rising in a horoscopal figure indicates a period of years and a repetition of this period in cyclic fashion furnishes a method of outlining the events likely to occur.

In the further consideration of the numbers and the possibility of utilizing them for prediction in events that will be in accord with the nature and quality of that number found upon the ascendant, as for instance, Libra would be upon the ascendant which corresponds to the number seven; then number one would be found on the seventh or thereabouts. If Capricorn be ascending, then Leo would be found on the seventh, corresponding to the tenth and fifth signs respectively, and the same application can be made to all the houses contained in the figure.

In judging the character and general fortune, this is not so difficult. In considering the date of events, while in general principle possible from the association of the numbers with the signs of the Zodiac and planetary rulers, when we descend into the detail, the complexity becomes great, for not only is it necessary to sub-divide the periods into successive years of life, but there is, in all cases, more than one cyclic period in operation during the physical expression, and this is in accordance with experience otherwise received, in the same manner as other parts of the horoscope besides the ascendant and first house have to be considered.

In giving an adequate account of a map, it is only reasonable to consider that other cycles besides that indicated by the rising sign must be of the greatest importance, and their values carefully weighed and interpreted. In fact, it must be kept in mind that all these cycles are in operation at one and the same time, beginning not only from the moment of physical birth, but from the time and moment of conception, and these are all related to other cyclic periods prior to these, that are incomprehensible to the finite mind, but their influence follows the ego in its flight in all its varied expressions.

Thus the student may realize what great cycles of time must be considered in association with any individual, and how much there is outside of the physical chart that is not contained therein, but which is all connected through exact planetary laws that are in operation just as effectively as those of the physical chart; but it is the task of the student to prognosticate and interpret and ascertain which of them are of the greatest importance in any horoscope and then to apply them properly.

We shall consider the association of these cyclic periods and numbers with regard to showing the events in life in the near future when we have entered a subject that will better enable the student to comprehend them fully, or at least sufficiently to utilize them in a practical way and demonstrate their existence, as we have sought to do in the various subjects that have been introduced. Our desire is to make clear every thought as we come to it. However, we have not yet arrived at a point where we can take up this subject conclusively and make it plain and simple to the mind of the student. This we can do as there are an infinite number of thoughts to be conveyed as you will have realized ere this.

We will now sum up the total of our considerations in the following manner: Nought signifies Parabrahman;

the monad or one, duad or two, and triad or three must be considered together and constitute the divine trinity. Then comes number four or the tetrad, the spiritual root of matter undifferentiated. The matter of our earth planet was inherited from the preceding lunar chain and of which we find the characteristic number was four, and thus we see the correspondence between the tetrad or four and the sign Cancer, the fourth Zodiacal sign and its ruler.

The Moon is very close indeed and none have seen its manifestation and application as converted into theological teachings as the virgin mother, the mundane egg. Then we have the signs, Leo, Virgo and Libra, that is the number five, six and seven, signifying the three Nirvanic planes and their correspondences; the number eight or ogdoad, signifying the spirit matter brought down to a lower stage and synthesizing the four lower planes; number nine, the Buddhic plane; the decad signifying the mental plane, the human ego, and the hendecad or number eleven signifying the astral plane, and the duadecad or number twelve signifying both the physical plane, also the celestial plane beyond finite conception, expressed in the sign Pisces.

CHAPTER XIV

The Awakening of the Soul.

The many stages through which the soul passes as regards ingrowth and development are often unnoticed by the individual as they seem to be lived in some unfathomed deep of his nature and he may not be conscious of the change until an incident or circumstance in his outer life reveals it to him, and the veriest every day trifle may be sufficient to pierce the veil and transfer to the lower mind that more complete and full consciousness which has been slowly builded through many ages.

It may sometimes occur that the individual is called upon to choose between some trifling enjoyment and some equally small act of duty. The pleasure insists as it has so often insisted before, but the individual hesitates until suddenly there arises in him a vivid consciousness, a sense of imperious self-control in almost ludicrous contrast to the magnitude of the decision it enforces, and this constitutes the awakening of the soul. True, this does not cover the experience in all cases, but it is an instance.

Then again it will be found that quite frequently it will be by the way of the emotions that this sudden awakening will take place. Thus we can realize the psychic qualities possessed by so many who come under the dominant influence of the watery triplicity. Those whose emotions are strong and are easily moved by their environments would come to a realization of self enlightenment more quickly and respond to the call. This awakening may come through emotion caused by the profound and silent depths stirred by the advent of a great joy, or through the sacred touch of a great sorrow, or the sense of melting pity moved by the sight of stricken age or the mysterious tenderness felt at the birth of a babe, and while the circumstance may be nothing, the revelation is everything, for it is the advent of the larger, higher nature, and the little arc of earth life seems to have been extended in order to embrace its vast and hidden circuit, and the individual has, so to speak, in a few moments, crossed an abyss of time and he becomes in a measure separated from his past expression, his former environments become distasteful, and at this stage the soul is the soul of a reformer, for having entered a wider realm of thought it naturally desires that the exterior conditions shall harmonize therewith, and the old dominant idea that the interests of the many must be sacrificed to those of the unit, falls before the recognition that the welfare of individual must sub-serve the progress of the race, and thus energy will be thrown into those movements whose aim is to make pure the impurities that exist about him, and he seeks to beautify the conditions of earthly existence and to introduce freedom, health and happiness into the collective national life, and it is through this awakened spirit that we may expect to realize a condition of peace and universal brotherhood of man on the physical plane of life, whose influence extends on to the other planes of life where aid is essential in bringing the wandering soul, who, steeped in the ignorance of undeveloped good, is seeking to find light, but seems lost in the dark night that surrounds him on all sides, the result of errors on the earth plane.

All these efforts on the part of the individual who had become awakened, imply a certain general practice of virtue on the part of the individual, and there will even be a marked growth in the more lofty emotions of the soul, when nature may open out to exquisite perceptions of color and harmony.

It may see, where coarser eyes are blind, hidden

beauties in the flower and forest, and may feel a more sacred joy in its friendships and a deeper sadness in its griefs than the ordinary individual, but no matter how moral the individual may be, even though super-moral when compared to those about him, and no matter how wide the range of his sensibilities or how complex his emotions, some change must take place in the nature ere it can be said to be definitely leading the higher life, and that is the recognition of the purpose of being, and its accomplishment through devotion of service, merely a change in consciousness, an attitude of the mind, but until this takes place there is no conscious, deliberate attempt to tread the path in the sense that is implied.

Unfoldment and Development of the Soul.

There may be a blind seeking for the light through action, but not a studious co-operation with the law which makes for righteousness, an endeavor to adust our complex nature to the conditions of its growth.

What, then, we may ask, is the purpose of life, or man's destiny. The goal of the victorious soul is a God-like condition, the Logos to rise triumphant over all imperfection and in the amplitude of time, with the song of eternal life upon its lips, merge its transcendant beauty in the same perfect condition with the all-infinite intelligences, and when once this mighty truth has been realized it will introduce a most powerful influence into the life, the nobler and better qualities will be assiduously cultivated, as purity of life, love, tenderness and compassion, selflessness, wisdom and joy.

The inner attitude will always be one of devotion and sacrifice even though the outer world duty is performed to kith and kin, to friend and foe, and there is earned the wherewithal to feed and clothe the physical body.

The nature will be taken in hand with scientific carefulness and there will be gradually but steadily developed a larger measure of hope, cheerfulness, tolerance, sympathy, love and confidence.

Then, too, the aspirant to this higher unfoldment will of necessity be compelled to cast out every shadow of desire, and in the performance of a service he will endeavor to find his joy in the pleasure that is given, or reap a loftier happiness in the bliss of having expressed the nobler life which is the life of the Gods. Then this act will be performed solely for that other, himself being but the instrument; nor wil. ne hold back whatever from the full measure of the gift in order that he may in secret offer it at the shrine of his own personality, and there is nought that will stop the free flow of the perfect love through him as a channel of the divine life.

True, we must consider the fact that there are many stages on this pathway, and they who but tread the green slopes at the base of the mountain on whose summit dwells the eternal snows of purity, cannot realize to the full the sublimity of those divine heights and it should be the care of the individual to trust that loftier self which mirrors the threefold radiance of God made manifest, and to believe always that the preponderance of power rests with it.

Then will be developed in the soul what we have considered a necessity to the world's unfoldment, that higher consciousness against which incidents and happenings that would disturb the serenity of the individual fling themselves in vain effort to disturb its tranquility and harmonious being, for the individual is awakened, and has created a power within and without that nought can enter except a power in harmony with perfection that exists with him; and these incidents and happenings that would disturb the ordinary individual have no more effect upon him than if it were a butterfly beating its wings about his feet, thus receiving events with lofty mind.

The life of the individual becomes infinitely more filled with lessons, and every circumstance is valuable in so far as it enables the soul to respond and grow; while the life of many who wander aimlessly hither and thither will be saddened or lightened by the thing that may chance to befall them, but with the individual who has attained the consciousness of being and has scaled the loftier heights of wisdom, it makes no difference what may occur.

The soul lights up all about it, all is illuminated by the true inward life, and for instance, the individual has been deceived, it is not the deception that matters but the forgiveness to which it gave birth in the soul, and the loftiness, wisdom and completeness of this forgiveness, and by these shall the life be steered, by these shall the eyes see more clearly than if all men had ever been faithful and true, and when love manifests itself, it is not the love that forms part of the individual's destiny but the knowledge of self that has been found back of that love.

In dwelling upon the higher truths of life and the esoteric, we find a greater field for expression, the only difficulty is that the material expression is so limited to convey to the mortal mind all that would give light, and thus the necessity of these lessons, daily happenings and circumstances that come and go in the life of the individual; but though they may pass on like the waves of the ocean they have left their impression according to their power and volume and according to the condition of the object they have met with. Thus we may realize how life is lightened and saddened by these happenings that are daily befalling them, these obstacles that are constantly appearing in their pathway as a barrier, between the individual and the goal he is seeking to gain, and at times petty circumstances will upset him entirely rather than affairs that are really more important to his well being.

But the point for one to consider above all else is that it is not the qualities of mind that will lift up the individual to a realization of conscious being, it is the actual knowledge back of the mind, and thus the mind is only used as an agent by the soul to convey wisdom to the lower consciousness, and thus will illuminate the material objective and lift it up to realization. It is the realization of these seemingly small occurrences in life that are necessary to attainment of the wisdom and power that is only latent and waiting opportunities, as for instance, one has found his confidence misplaced in one of his fellowmen, he has gained thereby, and if by this act of deceit upon the part of him that he trusted, the individual has not attained a wider range to his love and there has not come simpleness and loftier faith, then he has been deceived in vain, and the experience has been for naught; and it is the individual who is engrossed in the material and sees naught in life but matter, whose love for his fellowman becomes colder and more unforgiving, though if he would but consider that he is binding the chains of fate more closely about himself he might awaken to the truth that bitterness of thought only turns to evil to himself.

What indeed can stand in the path of a soul who wrings sweetness and beauty from events and circumstances that men call evil? But if the individual will but pause to study well these numerous manifestations that seem to him crosses, he will realize that this is the true scientific creation of the building of character, and which is followed by the awakened individuality.

It is along these lines of expression that the true disciple of the higher life builds his character and destiny, for he has realized clearly that the development of faculty is the anticipation of destiny, and the strength he gains today will eradicate the sorrows of the morrow, and the hope and joy that he attains today will give him strength of purpose to defeat the darkened fate that seeks to find and hold him at every turn.

Thus the only condition or the principle one is to open

the soul to receive, though it will be found that this revelation and training of self is but the prelude to sacrifice of an ever-deepening nature, and every lesson learned, each victory won, every step that is taken will impose its own obligation upon the individual. Its ever-growing compassion binds the ripening intelligence to the service of its fellowman, so that in reality by just the greater spiritual heights his expanding soul soars above his fellowman, in the same measure will he be chained to willing sacrifice on the earth plane, for the law of the highest life is the law of love, and as its devotees attain, they pour out to all humanity from their inner sources of being the glorious life that flows through themselves.

Thus from the mire of earth to the glory of the divine and from the selfishness that holds to the selflessness that outpours of its very essence, the soul climbs the great path, which though beginning in darkness finds its goal in the light of wisdom.

CHAPTER XV

Objective and Subjective Concentration.

We desire to dwell, for a time, upon conditions of life as they exist in a general way as regards the development of science of life. We find in the material thought realms emanating from mortals, there is an awakening to the needs of a more complete understanding as regards true science, and greater efforts are being made here and there. We observe results. There is a more careful analysis and investigation of forms by students with the purpose of discovering the life that lies within the form, and it is interesting to observe the methods pursued to arrive at conclusions, and come in touch with the unseen forces so termed, both in the realms of spirit and that life that is finding expression through matter forms. All seem desirous of finding the truth, though the methods pursued in order to obtain this knowledge are often diametrically opposite in their nature.

It has been said and taught that he who would pierce the veil and come in touch with the life back of the form must first realize and comprehend the life within himself, though it is often the case that for the purpose of analysis and in the attempt to penetrate Nature's secrets, the attention is chiefly directed toward the exoteric.

Thus the most fine mechanical apparatus, becoming more essential and more complex, intricate and delicate instruments become seemingly necessary, in order to accentuate the senses of sight, hearing and touch, and these outward appliances seem to these individuals to be the only means by which he may come to an understanding of these scientific truths. Then again, there is occasionally one who adopts other methods. First of all, he considers that as a fundamental principle, life and consciousness are one and the same, that these terms are synonymous, and that if the individual will study and endeavor to unfold this consciousness within himself he will become responsive and receptive to the finer and more delicate vibrations, and thus be able to receive knowledge by direct perception and receptivity, and in this investigation there is much work and study, but at the same time he is unfolding his own consciousness, his study is himself, and he seeks by careful training of the thought by method, discipline and general control of the lower self, to bring to the surface from the inner depths of his own inner being, the powers and forces that are within. Just as in the tiny seed lie hidden all the possibilities of food and nourishment for the physical man, or the great tree under whose spreading branches man may rest from his daily toil, it is all there, waiting unfoldment, and so hidden within each human form is this life of the divine one aspect of which is wisdom, the power to know by direct cognition. Then again, the individual who seeks to study the life from within applies science to life rather than to form, and this constitutes concentration and meditation.

Concentration is taken from the Latin *con,* interpreted implies with, and *centrum* meaning the centre, the full interpretation, with the centre. This term concentration may be divided into two parts, that is, external and internal, objective and subjective. Objective concentration gives the individual a perception and knowledge of differences, while subjective gives one a perception and knowledge of likenesses and similarities.

In the objective, concentration, an effort is made to draw the mind to one definite thing, and force it to hold an idea or subject in much the same manner as the hand is made to grasp an object and hold it so long as the individual wills. Thus in one instance the thing is taken up for investigation from an external point of view, while in the other instance the mind takes a subject for investigation from the esoteric, though in both instances the object or subject must be held for a given length of time before any definite knowledge can be gathered from it, and in the daily life of the individual many opportunities are placed before him for the cultivation of the objective concentration, and no matter whether the individual realizes it or not, he is cultivating these qualities whenever he seeks to become the master of his work, no matter what that work may be, or however relatively unimportant the task imposed upon him.

The shoemaker who makes a pair of shoes perfectly has become an expert along that line of work. The mechanic who can build a perfect machine and operate the same; the carpenter who can build a house that will withstand the wintry blasts, all these become masters in their particular line of work, and have consciously or unconsciously cultivated objective concentration, but just at this particular point in the unfoldment the individual must face a more or less danger point, that is, in becoming too much absorbed in the object upon which he is concentrating, as for instance, the individual who is entirely devoted to the acquisition of wealth, fleeting honors and fame, at last becomes totally absorbed by that object which usurps the place of master, while the individual becomes the slave, and even should not this condition arise where is slave instead of the master, there ever arises the danger that may be so often observed, that is, the danger of unequal development, as the mind, ever being permitted to rush outward without careful thought or consideration, and subjective concentration being practiced in order to gain and hold the proper centralization, balance and harmony.

Subjective Concentration Gives the Inner Life an Opportunity to Manifest Itself, Thus Gaining Power and Ability in Action.

This subjective concentration gives the inner life, the soul of the individual an opportunity to manifest itself, and through this means will gain power and ability in action in the objective or matter world through the force gathered in the subjective plane and sphere of life.

On the side of the objective concentration we have reasoning, deliberation, forethought, calculation and practical common sense, while on the side of subjective concentration we have calm passivity of thought, and the mind becomes receptive and permits the knowledge that lies within the soul to become manifest to the lower consciousness, and by its being impressed upon the brain centre it naturally becomes of use and lasting service to the individual, and creates an environment about himself that impresses and affects those who may come in contact with him according to their sensitiveness and receptivity.

Thus we may realize the results when one individual seeks to live in this higher expression of life. He will have an influence for good in the community in which he resides, and be of service to himself and others, though by investigation it will be found just as difficult to cease all mental activity at will, as to compel the mind to hold to any given subject for any given length of time, and it is well that this is so, though the mind will be found to be the only true instrument of investigation and research in the science of life divine, and until man has become master of that mind the secrets of life must remain more or less hidden from his view.

When we realize that the mind is the only true instrument that may be used in the investigation of the science of life, we may understand to what an extent thought enters into the problem of life's unfoldment, and all the life is moulded to a certain extent by thought, and thus we become our own creators while Omnipotence acts through its own physical manifestations by the power of thought, and every conceivable form of life has a desire to improve itself and its own environment, and this comes under the law of love; though it must be understood that the result of such desire is not always manifested to any great extent in that life expression but on another, higher plane of life, and herein lies one of the truths of the indestructiblity of life.

As we learn it is not destroyed by its constant change of expression, but it must go on to the goal of perfection, and we may realize how these qualities of concentration and meditation affect these various expressions of life and change the personality, and in a careful study of the animal life we can observe the power of thought, as for instance, some insects and animals are enabled to conform their color and appearance to their surroundings in order to give them a fairer chance against innumerable enemies, and on the other hand to enable them to more readily approach unperceived their natural prey, and by this fair division of possibilities is the balance of Nature's law maintained, though how often do we have to deplore the disastrous results of man's interference with this balance, and it is thus we find the necessity of concentration and meditation on the part of humanity. We have considered two kinds of concentration, the objective and subjective, though many individuals are quite apt to confuse concentration with meditation. While concentration is a centralization of the mind or thought, meditation is drawing to the centre of one's own being, and from that centre being able to direct the mind and guide the senses of the physical form.

We have the teaching, the kingdom of heaven is within, still we may find many today regardless of this teaching seeking the kingdom in all directions, looking everywhere but within their own being, for the peace that passeth all understanding is not to be found in the objec-

tive world, and can only be attained when emotion and
external thought is dropped, and a higher state of consciousness has arisen wherein is reflected the peace of the
infinite even as the tiny dewdrop mirrors and reflects the
glorious sun.

Meditation is the door to the kingdom within the
individual and it must be kept in mind that at the centre
of one's being, the soul, the individual is no longer identified with any modification of the mind expression for it
is there alone, where mind itself is quiescent, can peace
truly be found and realized, and there are but a small
number of individuals in the mortal today who are truly
able to practice meditation and the reason of this is quite
clear. All their mental images gather around them and
bar the way to the centre, and the demon of doubt is one
of these images, which when left to his own freedom has
the power to weaken and retard the action of the will, and
in an astrological sense we have found Saturn to be the
planet that has chief control over the individual with
regard to meditation, and thus while his lower influence
is unpleasant, disagreeable and even severe, his true
higher influence is most essential. He is the planet of
steadfastness. His vibrations produce profound thought,
earnestness, patience and a desire to know. He is the
great pain producer and the reaper of experience. Saturn
afflicted is considered to bring sorrow and pain, though
sorrow exists for the purpose of driving the individual
from the external to the centre of being, and Saturn is
really the planet Jupiter reversed in his influence, and
the experience of sorrow is essential, in order to compel
the soul to find its own peace within, and through this
peace comes a realization of unity and bliss.

CHAPTER XVI

The Aura of the Human Form.

Every human being is surrounded by a force, a vibration peculiar to that individual alone, that is a luminous mist, called the aura.

Those individuals who have developed the power of clairvoyant sight are able to perceive these finer vibrations, and are thus able to describe them in the form of colors as they appear to them. These vibrations ever harmonize with the quality of the thought, character and individuality of the person. Naturally the colors become more refined and clear as the soul unfolds through the process heretofore described, and find by investigation that every object of life and form possesses an aura or atmosphere peculiar to itself, all things in the vegetable, mineral and animal kingdom, though it would naturally be less complex in quality than with the human soul.

Now the life principle that is being poured out from the planets, and more particularly the Sun, is specialized through the organism of the individual, and the spleen is considered to be the material organ through which the life force is manifested, and that the spleen comes under the rule of Saturn; thus the stronger and more robust the constitution of the individual the more of this life principle they will receive; and here, too, may be seen the result of an afflicted Sun and Saturn in the horoscopal figure of the native, as this affliction naturally interferes with the harmonious flow of this life-giving force, and on the other hand when both the Sun and Saturn are unafflicted in the figure, the native will receive and radiate a greater amount of the life force, as for instance, here and there is found an individual who can give off an enormous quantity of this life force through magnetic contact with another who is receptive to such forces, and such individuals are constantly giving out forces to those about them whether they are conscious of it or not. On the other hand the individual who is not able to specialize a sufficient amount of these life-giving forces for their own needs will unconsciously, perhaps, absorb and draw from those about them that are in sufficient harmony and receptivity for them to draw from, and in this way live to a great degree upon the life forces of others with whom they are thrown in contact, and then, too, as the form advances in age this force does not specialize and radiate through the organism, and thus we find the physical form becomes withered and bent through a lack of this life principle, and thus we find that an elderly person is quite apt to draw from a younger person through whose organization these forces radiate more freely, though there are exceptions to this rule, as much will depend upon the positive or negative nature of the individual, and many have no doubt felt drawn out, as it were, after coming in contact with certain individuals.

Then again, on the other hand, there are instances where these forces are exchanged with benefit to all concerned. These conditions spoken of are often witnessed in a seance room where a number are gathered together, and unless the guides of the circle understand this law some will retire from the seance room quite exhausted and still others are not affected.

Now the first misty-like outline that may be discerned by one sufficiently clairvoyant to perceive it is termed the health aura and is generally visible as a mass of faintly luminous bluish grey mist, often tinged with orange or gold, though often this aura is colorless and seems to be formed of innumerable lines in a perfect spiral shape surrounding the physical form and radiating evenly on all sides, that is, where good health prevails; but in ill health the lines are irregular and imperfect, and then the health auro is weak and the spleen cannot specialize sufficient life force.

We may by investigation find that the causes of this disturbance in the health aura, and which has a direct influence upon the harmonious vibrations of the human organism, are many. One of the most common is a tendency to think and act along one groove. The energies are confined to one special channel and the nerve forces that are played upon become exhausted and then comes affliction, and the cause is often mis-named overwork. Then an extreme depression of spirits will cause ill health and retard the natural inflow of the life principle from without to the form physical. Worry and melancholy are great factors in disturbing the natural flow of these vibrations and are often the cause of ill-health, for this is where the mind and thought has power to affect the conditions of health, and by keeping the thoughts pure and the mind free from worry and dire forebodings, the health will take care of itself. The point in view is not to obstruct the way for the harmonious inflow and outflow of these life-giving forces.

Often a sudden shock will throw the current out of its natural channel or a wound or injury will have the same effect for the time being, but once the mind can become passive and gain control, the nerve forces begin to regain their natural condition and the obstructions to the inflow of the life principle are removed and all is in tune once more. It is at such times that this inflow of force is obstructed, that the contact of another individual who is overflowing with this vital life force will set right the imperfect operations that are the cause of sickness and affliction to the physical.

We now pass on from the health aura and come to the animal desire. This is controlled, as we have learned, by the planet Mars and the Moon, in the same manner as the health aura is ruled by the Sun; and the desires of the individual are manifested through this aura, which is very changeable in its color and vibration. With the majority of individuals who are living on the lower planes of life this aura is highly colored by the Mars influence, scarlet and red predominating. The natural aura that surrounds the physical frame is rather complex in form, color and vibration, and to one who is clear sighted it may be seen as a luminous cloud extending to varied distances from the physical body, from two inches to two and three feet. This aura is generally oval in form and in ancient writings is referred to as the auric egg. In the majority of cases this aura has no well defined outline but the outer edges gradually fade into invisible vibrations.

A close study will show that it possesses several distinct component parts, and that these several component parts are really matter in different states of vibration and that each of them appears to be distinct from the other, in fact, a separate aura, and it would seem that if the others were withdrawn to occupy the entire space as the entire aura itself.

When the individual has attained the higher understanding of life in the mortal or spirit, and really lives in accord with the same, there will be seen seven of these seemingly distinct auras, while in many instances there may appear to exist only one or two, as will be readily seen, depending upon the extent of the soul expression from the within, or whether it is buried up in the animal nature. In viewing this aura, the individual who can see and discern finer vibrations than material, it is possible for him to perceive up to five; beyond this he cannot see, that is, the individual who is manifesting in the mortal life.

From an astrological point of view, man may be considered as a star, radiating in space, with seven concentric, and the mortal, invisible rings or belts governed by the seven planets, while the life principle of the whole is the Sun.

Descriptive of the Different Auras, the Effects of Thoughts and Emotions Visible to the Clairvoyant.

In a further consideration of the aura of the human form and the soul itself, we have found that of the seven belts or rings of atmosphere the lowest and most material is the one associated with the physical or matter form and which is termed the health aura, and it will be found by those who are able to discern, that the general coloring and appearance with regard to form is found to harmonize with the healthful condition of the physical body.

The next separate aura is the magnetic, although in a sense the magnetic and health auras may be considered as one, as they are mutually interactive at times and the conditions of the one act upon the other.

Then comes the desire aura, that comes under the domination of the planet Mars, and to the individual who can discern these vibrations, this is the mirror that reflects each and every desire of the native.

The emotions and feelings are expressed in this aura in the great majority of cases, and a major portion of the thoughts are manifested and find expression in this desire aura, which is more readily perceived by the clairvoyant than the other auras, and from it are sent out thoughts for good or evil which are created and set in motion by the will in expressing the desires and feelings.

When good thoughts are sent out in a purely unselfish manner, good is the result. When gratitude and love go forth, good results therefrom, sending beneficent vibrations to humanity that assist the onward progress of the world in the unfoldment of the soul forces.

On the other hand, thoughts of evil and wrong to others have the opposite results, and all such low order of the animal nature such as maliciousness, anger, jealousy, envy and spite are the seeds from which are harvested sorrow, discontent, unhappiness, pain and affliction, to say nothing of the results in retarding the wheels of progress in the world's evolution. All this enters very closely into the study and interpretation of Astrology and explains very aptly how the chains of fate are forged about the native, binding him closely to the sorrows and unrealities of life. It gives the explanation of how the individual can and does create his own destiny and demonstrates that he alone is responsible for his spiritual and physical condition.

We have already learned that as we think so likewise shall we become, and as we create character and individuality by the desires and thoughts, in a like manner do the stars and planets reflect an exact likeness of the native in the horoscopal figure.

Thus will be seen the absolute necessity of cultivating only good thoughts, for although all vibrations of thought will finally be merged into good by their contact with the good sent out constantly by those who are living in the higher understanding and true wisdom, both in mortal and spirit, the evil thought does its mischief and is so much more poison to be purified and cleansed of its impurities; thus the law finally masters but the pain and suffering cannot be calculated by the finite mind, but the lesson must be gained no matter how bitter the experience nor what the cost may be. All thought sent out finds existence according to the will back of the thought and the wide spreading results of its activities during the time it is in existence. Thus again we see the truth expressed, as a man sows so likewise shall he reap.

Next to this desire aura and closely linked with it is the fourth belt, which we may term the thought or mind aura and from this may be described the record or progress of the personality.

There is nothing of so much importance to the individual as a general understanding of his environments, as it will permit him to overcome and master the fate that seeks to bind him to matter.

In a further consideration of the mind or thought aura, it may be said to reflect to the soul the general nature of the desire aura that is outside of it, and then again the desire aura may be said to reflect the general nature of the thought or mind aura, though it must be understood that there is much more to the thought aura than to the desire aura, as through the thought aura there sometimes appear flashes of intellectuality and spirituality as the soul seeks to find expression, and these qualities do not harmonize with the desire aura, that is, where the intellect is in its proper balance, and at the same time, if the vibrations of consciousness corresponding with any particular desire are repeated strong enough and continued for a given length of time, it will naturally react upon the mind and will set in motion corresponding vibrations therein and thus tinge the thought aura for the time. If an individual was on the line of evolution and progress through good thoughts and wishes for all humanity, continually sending out aspirations towards the infinite good and with a sincere desire to aid his fellowman, the thought aura would then take on the colored vibrations in harmony with the desire, the color being of a light blue tinged with a very light yellow, and if the desire was of a personal character, a sincere love and affection given out for some dear friend who was in distress, then the color would be of a light blue and very light yellow tinged with pink, and so on through the various expressions of the thought and mind.

In this thought aura may be observed both the good

and evil emanations of color vibrations, or rather that which may better be described as undeveloped good. This is the result of ignorance of the law and humanity must come to a realization sooner or later that so long as they permit themselves to act as an instrument for low, undeveloped, imperfect thoughts, just so long will they remain in vibrations of inharmony and discord, the imperfections becoming intensified and the scarlet colors of the animal nature, the black clouds of melancholy and depression will assail the good within that is awaiting opportunities for expression through these same auras. Therefore let the mind and thought act in perfect harmony with good if good and pure environments are desired.

We next come to the aura which we may term the lower soul, the aura that is associated with the matter form and the soul. This is the fifth aura and many mistake this for the real soul aura, but this is not true, for as we have stated, the mortal eyes cannot behold the sixth and seventh auras no matter how clear sighted. There are none who are finding expression in the mortal at this present time. There may come a time in the world's evolution, without question there will, when those in mortal expression will perceive and know more than those today. This aura expresses the individuality in the world of matter, and we have learned heretofore that the individuality is symbolized by the Sun, while the personality is expressed by the Moon, the one being permanent and lasting, while the other is fleeting and changeable. The color vibrations of this aura are of a more delicate and refined nature than the other auras referred to, and as in the other emanations of mind and thought, it depends upon the expression of good that dwells within for these fine, delicate colored vibrations that should be found here. In instances where the individual is living in the higher understanding these vibrations of color are almost of a crystal transparency, or at least they appear so to one in mortal expression and is naturally quite difficult to describe.

Aura Begins to Clear as the Soul is Awakened and the Mind Begins to Think Along the Lines of Spiritual Things.

We have considered, in a general way, the five auras that surround the individual, and it is not difficult to realize what a vast field for study and observation this subject affords, and those individuals who are clear-sighted enough to discern these auras can better appreciate the varieties of colors that vibrate and go to make up the environments of the individual, for unless the individual is manifesting in harmony with his aura, or in other words, if he is not surrounded by conditions that harmonize with his aura he will not feel at home and will be out of place elsewhere.

As time goes on and the spiritual sight becomes more clearly developed it will then be an easy matter to know individuals as they really are and not be confused by the mere objective physical appearances, and those who are willing to live in harmony with the law and manifest in truth, purity and love and abandon the personal desire, with a sincere, earnest aspiration to serve humanity, such may realize more of these realities than others who merely live in the shadows and delusions of the matter world, and they will find that the inner sight will slowly but surely unfold, for it will be found that this spiritual sight is in the majority of instances one of the evidences of the opening of the soul sight, a higher sense of seeing that can be developed by every individual in time as well as that of clairaudence, and feeling more than is ordinarily heard or felt by those in the mortal expression, and even those in the spirit must cultivate these qualities if they would realize all that is about them, as it is often the case that a spirit comes to the higher plane with all the spiritual faculties dulled by his absorbing interest in the affairs of the matter world, and his position is not one to be envied though he fails to realize all that he is missing, for his aura has become so dense and colored that naught but imperfections can enter therein.

It is only as the soul is awakened and the mind begins to think along the lines of spiritual things that the aura begins to clear and the atmosphere is cleared so that he can perceive what is around and about him. Once this unfoldment is reached, the rest is not so difficult to attain. It may be well here to mention some of these aura colors and the cause thereof as given astrologically, as it will serve to make these lessons more clear to the student's comprehension. In instances where the colors of the aura are heavy, thick, black, and lead color, it is an indication of maliciousness and hatred, and denotes that a small amount of the principle of the vital life forces are being received by the individual. In such instances the ruling planet in the natal chart will be found to receive an affliction from Saturn, and the Sun and Moon are also apt to be found in affliction from the same source. When dark red flashes manifest, then anger is shown, the colors becoming lighter as anger lessens to indignation, becoming light red in color. In such instances Mars will be found near the ascendant or afflicting the ruler or Saturn. When a lurid red flame of color manifests itself in the aura of the individual, which is quite pronounced, then the animal passions are indicated, and Mars will be found afflicting the first house, or the Moon or Venus will be receiving his malevolent influence. When a dark red brown is shown on a dark background then avarice, self-desire is dominating, and in the natal figure Mercury or the ruling planet is receiving an affliction from Mars. A dark, livid grey on a dark background shows fear, weakness, low vitality, and in the chart, Saturn and Mars will be found weak and afflicting the ruling planet.

In consideration of the colors that are to be found in the environments of different individuals in harmony with the planetary influences that are to be found in the horoscopal figure, one of the most common colors or combination of colors to be found in the aura, is a dark heavy brown mixed with a leaden gray. This vibration indicates that self is ever to the fore, selfishness predominates the every thought and action of the native. This is a result of Saturn afflicting the ascendant or planets found therein, or when Mercury and sometimes the planet Venus is receiving Saturn's evil rays. When a dark, heavy, leaden gray is observed in the auric colors this is an indication that melancholy and depression exists in the environment. The individual is ever looking on the dark side of life and will make much of a very small matter. When the color mentioned is mixed with dark green that may be compared to a quagmire, it is an indication that the individual is deceitful, dishonest, apt to take that which does not belong to them, is a gossip and mischief maker. This is when Saturn is weak and in conjunction with or afflicting the ascendant or Mercury and the Moon. Unfortunately these colors may be too frequently observed in the environments of many, and it is well to keep in mind the fact that those loved friends who have joined the great majority in the higher life must necessarily contact these varied imperfect conditions whenever they attempt to commune with them or come enrapport, and thus it may be seen how essential it is to eliminate all imperfect environments and vibrations, when entering the seance room for the purpose of holding converse with those in spirit life. It solves the problem of receiving perfect results. Then again, there are times when one individual will disturb the harmony of a gathering by the great activity of their own imperfect vibrations when contacting the vibrations of others who have placed themselves in a passive receptive mood, and naturally are more quickly affected by such influences than when out in the world of activity. Great attention must be given to these laws and an effort must be made to conform to them if the proper conditions are to be given to those intelligences who are seeking to find an open door to express themselves in perfect harmony. When a dark yellow is observed, tinged with green and occasional flashes of red on a dark background is shown, this is an indication that jealousy holds sway in the environment of the native, and his life will indeed be an unhappy one, not alone affecting his own but others' happiness, and turning joy to sorrow. In the natal chart Mercury and Venus are afflicted by Saturn or Mars, and even Urania or the ascendant is receiving the evil rays from some of these combinations of planetary afflictions. It is a most serious affliction indeed, and the cause of much sorrow and unhappiness in the world. All these afflictions may be intensified or lessened in their power by the individual's own will if he will only strive to know his environment and will act accordingly, thus changing these dark heavy color vibrations to the milder, lighter, finer colors. When a dark red is seen almost concealed by black, then there is an indication of rash anger, hatred for everything about them, never forget an injury, and will seek revenge for seeming wrongs done them. This is a terrible influence to find, for it denotes an individual that will, if this vibration is permitted to exist, commit some terrible crime, and when angered will not hesitate to take the life of a fellowman. Such conditions are to be found when Saturn, Mars and Urania are afflicting the first house or the luminaries, and Mercury.

Yellow Color Vibrations Indicate Love, Orange Color Indicates Ambition and Purple Indicates Intellectuality.

The yellow vibrations indicate love and are often a clear, beautiful color, not tinged with other colors, though naturally will be found to vary just according to the nature of the love that is manifested, being a light clear yellow or tinged with the crimson and red vibrations of passion, which is to often mistaken for love itself. However, if the yellow is tinged with the lilac or blue colors it indicates the love is of a higher spiritual nature and has a good feeling for all humanity, for the individual has then unfolded to that wisdom that knows all are one in the infinite life eternal. It will be observed that these colors range from the rose color up through the dark yellow into the lighter shades·of yellow, mingled with the violet and light purple hues.

Orange color indicates ambition, and according to the shade shows whether it is for self or for many. If it is dark and tinged with dark brown it is an indication that the native has only his own personal interests to further, and if dark red is interblended he will be cruel, unjust, and shows a tyrannical nature who will not hesitate at anything in order to gain his ends. When orange is observed tinged with indigo it denotes the native is proud and haughty. However, the light shade of orange prevailing with the lighter shades of blue, purple and yellow interblended, shows an ambitious nature in position of power and authority, who has the interests of his fellowman at heart, and is able to do great good in the world. The difficulty is that it is hard for the masses to recognize in the native one that can aid them, for the reason that the dark heavy colors are too frequently found in the aura of the majority, and the lighter colors heretofore described are too infrequent in their manifestation. It will be observed that in this aura the colors are many and varied in their expression, and it will require close application to the subject to judge of the nature as expressed by his colored aura.

The purple colors indicate intellectuality, though they will often be observed to run into the dark brown, yellow and red, which is an indication that the intellect, the mind and thought is being directed along the lower channels of life, and it should be the effort of the individual to lift up the thought, for herein lies the key to unfoldment of the soul forces within, and we see how Mercury is the messenger of the Gods, for through his expression in the natal chart is the mentality controlled in a great measure. When the heavy colors predominate we know that Mercury is receiving the affliction of the malefics, and that the native is making no effort to change or rule this aspect. On the other hand, when the lighter heliotrope, lilac colors are found with the lighter shades of yellow, blue and gold, it is a sure indication that Mercury is receiving the benefic influence of other planets and that the first house is not afflicted by malefic influences.

Among the first colors described, yellow, when the shades are light, indicates that Venus is in good aspect with the luminaries or the planet Mercury, and also that the first house or ascendant is receiving the benign influence, and Mercury is strong in bringing out the good impulses of the native.

Then again, when the lighter orange and gold is observed, mingled with the lighter shades of yellow, blue, lilac, etc., it is an indication that the Sun is well placed in the figure and is receiving the benign influences from the Moon and benefics as well as Mercury, also the first house is unafflicted and strong. It may be observed how essential it is to study the mental forces in the horoscopal figure in order to give accurate judgment of the environment, and how great a part the planet Mercury and the higher Urania play in the progress of the individual, all of which is found expressed in the colored aura.

The great infinite idea is to create and bring back into itself again, and we find the will of infinity so organized that receiving all the seeds into itself from the cosmos and keeping them within itself, will make them manifest and their dissolving by chemical processes make all new again, and thus as a gardener all things that have been dissolved it taketh to itself, and gives them renewed life and power, for there is nothing to which no life is given, but creating and bringing into itself it gives all life, and is at the same time the centre of life, and in a sense its creator.

Continuing our subject concerning the auric colors as they may be found in the horoscope, when in the horoscopal figure we find Jupiter in the first house and unafflicted, or when Jupiter is in the ninth or has rule over that house, and posited in a sign of harmony, especially when in his essential dignity and in the benefic aspect with the luminaries and Mercury, there is manifested a deep feeling of veneration. It may not necessarily be of an orthodox nature or confined to a personal being though this is often the case when Mars or Saturn are in aspect, and when Jupiter is receiving no aspect from Urania or Neptune. Naturally, when the indigo colors are deep and tinged with gray brown it is a selfish motive back of the religious feelings that prompts the native to worship, and when the dark green is tinged with a leaden gray it shows deceit and hypocrisy, and the native uses his religion as a cloak to further his own schemes. These colors vary from the dark indigo to the violet and lilac and light blue, just according as the planet Jupiter is posited in the figure and the aspects it receives from the other planets. When the indigo color is shown tinged with dark green then the native is changeable in religious affairs and goes from one belief or faith to another. His environments will control this, depending upon those with whom he is brought in contact. This is when Jupiter and the Moon are in affliction. When the dark

blue colors are shown clearly without the black mingled
with the violet and light purple hues, it is an indication of
ingenuity, a studious inquiring nature, very apt in learning and possessing a very high ideality, a mild reserved
nature. This is when Saturn and Mercury are in good
aspect in the figure and are found well placed, and if the
light green is interblended the individual will be an inven-
tive genius. When the orange or gold is interblended it
shows a strong vitality with a most humane nature, one
who will do much for his fellowman without regard to
position or station in life, will be a friend to the common
people for he will recognize their inexperience and their
needs and the necessity of showing them the light, or
bringing truth where error exists, and when the luminous
lilac blue are shown interblended with the gold it indicates one who has unfolded to a high condition spiritually,
aspirations, unselfishness and devotion to duty. This is
when Jupiter, Urania and the luminaries are found in
good aspect with Neptune in the first, tenth or ninth
house. It will be seen that all these colors in the aura
are subject to infinite combinations, so that to read in
detail the indication of the influences is a most difficult
task and requires careful study and judgment, just as the
judgment of the natal chart requires careful forethought
and knowledge of this divine science, for diversity and
complexity are the chief features in material expression.

A student of life's environments and physical condition can readily perceive that it is naught but selfishness, ignorance of truth, and unbrotherliness which
makes of humanity a sea of sorrow formed of the tears
of mankind. The darkness of selfhood chains man to
the animal plane. Altruism and self-forgetfulness reveal
the divinity in men and life him onward to the super-
human stage, but it must be kept in mind that the birth
of new ideas and nobler eras is accompanied by storm
and stress, through which all should pass undismayed,
secure in the knowledge that the peace and unity of the

sixth sub-race await our efforts to prepare the way, and thus amid the clash of war and contention and social revolution which will be our lot to witness, we shall feel secure in the remembrance that from the storm and chaos engendered by the warring wills of men, the new life of peace, brotherhood and unity will emerge triumphant.

CHAPTER XVII

All Forms from the Fixed Stars and Suns Down to
the Most Minute Atom Have Their Own Rate of
Vibration, and the Great Master of Wisdom Never
Makes a Mistake When He Hears the Sound.

All forms from the great fixed stars and Suns down
to the most minute atom have their own rate of vibration, and each individual born into this physical expression has his or her own diminant and characteristic keynote, color and sound, these making up together with
the other vibratory notes the chord or mass vibration of
the aura. This aura, or as we may better term it, the
individual atmosphere, is made up of life atoms in which
many varieties and types of vibrations are latent. It is
true that in each individual, one particular rate of vibration is dominant, one class of these life atoms more immediately operative than others. In order to know these
rates of vibration in one's atmosphere and the chord, as
it were, to which the individual is keyed, then the self-
consciousness must be raised from the personal self to
that of the divine within self. These vibrations move
both to the centre from without and from the centre
outwards, that is, vibrations tending inward and other
vibrations tending outward, one being positive, the other
negative, and the centre between these two vibrations,
the mean between the positive and the negative, may be
termed the keynote of an individual, or what we could
best term in this science, the polarity of the individual.
We then find that each one is similar to a musical instrument, keyed to a certain chord and is emitting a certain
sound, and the great Master of Wisdom never makes a

mstake when He hears the sound. We may term it the life sonud of any person or thing, for it is the dominant note, and gives the key, so that it can be told just where the soul has attained to in its evolutionary progress by considering the rate of its vibration. These vibrations also manifest themselves on the physical plane in the case of male and female, in the pitch of the voice, which is an index to the life's tone of persons generally, character being expressed in a remarkable manner by the voice; as for instance, an individual born under the influence of Mars, the voice is sharp, commanding, positive, and if combined with Mercury, is shrill and metallic, while the Venus note or sound is pleasant, gentle and negative, while the Jupiterian will have a deep and powerful voice, yet harmonious withal. Very few, with the exception of the student of the divine law, realize how clearly the character is manifested by the sound produced. There are sounds innumerable that one has not heard and cannot comprehend until he unfolds himself to a harmonious chord with the music of the higher spheres, and can only then begin to realize the wonderful vibrations of each separate planet with regard to its numerical ratio. An exact prediction concerning the coming destiny of any nativity can be accurately given only when the dominant chord or note of the individual is known, together with the rate of vibration at which he or she is manifesting; some individuals being able to exhaust unfavorable influences and cross vibrations much more rapidly than others, though this would naturally depend to a great extent upon the experience of the soul working and manifesting within its self-made limitations, and would be especially true with those whose minds were turned toward the divine higher life unfolding the divine laws in Nature, and who sought diligently to co-operative with her, thus to dominate their lower animal nature and in this way realize their higher soul ideals, and using the will in conjunction with the planetary forces by aspiration and individual effort the higher notes of the scale of each planetary rate of vibration would be brought into action and thus hasten the progress.

The way is being prepared for the coming generations' evolutionary progress, and we are manifesting in harmonious accord with the needs of humanity, struggling in the darkness for what they know not of nor can they comprehend, seemingly at times having it in their grasp and then when attained find the valuelessness of it all, and is not what was most desired. It is the soul seeking to express itself, but owing to the crude instrument through which it is seeking to make known its wants the result is imperfect and unsatisfactory, and this cannot be changed until the individual will finally co-operate with the planetary laws and play upon the finer, higher notes of each planetary vibration; as for instance, we may consider the vibrations of Venus, which are finer and more rapid than those of Mars, and these higher, finer vibrations tend to neutralize and harmonize the slower, coarser Mars vibrations; thus, as humanity thinks and lives purely, the coarser matter of impurity is eradicated, as in manifesting love, the coarser matter which we know as hate, is cast out, thus constantly replacing the coarse and crude by the finer, always appealing to the centre of our being, the true self within. Let us consider more closely; as for instance, an individual comes into expression under the dominant influence of Venus, and that this planet's influence or vibration was affected in its action by the slower vibrations of Saturn, or as we would ter mit astrologically, afflicted by the planet Saturn, then we would find that until self has been dominated, the rate of progress would be very slow, and great suffering would be experienced by the individual until the soul had learned the truth that all are one, and in order to find progress must assist humanity along with him. Thus, the truth would then set him free, for it would have realized through sorrow and pain that separateness and self-seeking do not bring happiness and joy that are lasting and eternal, and the vibrations of the planet Venus would be liberated from those of Saturn, and evolution and unfoldment of the soul could go on unhindered, and the physical man which would be most affected by these vibrations could then provide a more sensitive and finer instrumentality for the soul's expression. It will ever be found that those persons that come into physical expression with the planet Venus free from the affliction of the malefics Uranic, Saturn and Mars are lovers of mankind, most unselfish souls, compassionate and tender, ever thinking of humanity before themselves, as the love of the solar logos conveyed through the vibrations of the higher Venus would thus be able to flow through the instrument of personal self unhampered and without obstruction. Again, if the planet Mercury, the ruler of the mind, is vibrating in unison with the benefics, as the Sun, Jupiter, Venus, then the manifestations of the self as the knower, will be enabled to use the lower physical as an instrument to express truth, the mind being lofty, generous and noble, able to study the great essence of life, to be profoundly meditative, contemplative and wise; but on the other hand, if the Mercurial vibrations be afflicted by the malefics, Saturn or Mars, then Mercury, the mental ruler and messenger of the gods, will be unable to transmit his message, his wings being smirched with the dress of the earth. He will become a captive, fettered by the bonds of self and sense, and the mental images manifested will ever be of the earth, earthy.

We have learned that an afflicted Mercury clouds the mind and will not permit a pure, wholesome manifestation of though as Mercury is the chief ruler of the mind, and when his vibrations are affected by the affliction of Saturn and Mars, the windows of the soul become dark-

ened and the messenger of the higher wisdom is held
captive. and the great solar light which lights the souls
of all humanity will be unable to pierce this density of
inharmonious vibrations which have been set up through
the natural planetary influences into which the individual has launched his boat. Those who have unfolded
the spiritual sight and can perceive the colors that surround the individual, will find that the lighter and finer
the colors are in the aura, the more pure and perfect
will that individual express himself, sathe soul will find
a harmonious pure instrument through which it can
manifest its divine power. It is a truth that each atom
that goes to make up the atmosphere of the individual
is created by certain rates of vibration which have their
color accordingly, and the heavier and darker the color
the slower will be the rate of vibration, and the lighter
the color the more rapid will be the vibrations. In the
consideration of the Venus influence which we recognize
as love, the higher octave has been reached, and the
whole gamut of the lower vibration of Mars has been
overcome, and the deep scarlet which is created from
passion, lust an danger, as found in the Mars vibration
is transmuted and replaced by the lighter, finer pinkish
tint that speaks of love and affection. The mind is found
to be turned more from the lower concrete when the
orange associated with Mercury is changed to a light
yellow as the mind will then manifest more unselfish
characteristics.

That which marks the higher evolution of man is the
breaking away from the barriers of selfishness and pride
that characterize the animal ego, and to overcome these
is to break the chains of bondage to escape from the
fetters that bind, as it were, from the chrysalis state, and
unfold the higher spiritual. It is to pass on from the
narrow plane of the personal into the birth of the universal; and if we will look for a moment at this physical
world with its most marvelous discoveries, seemingly, its

progress in all the fields of science, invention and art, we shall perceive that during the Sun's progress through the Zodiacal sign Pisces, the watery sign, steam was the greatest factor in all the many improvements and inventions, revolutionizing all the industries and bringing the various countries and nations closer together through greater facilities of communication. In a similar way only, upon a higher round of the ladder, will the Sun's passage through the Zodiacal sign Asuarius be marked by discoveries and inventions relating to the air and ether as the agency brought into operation; as for instance, the communications without visible substance as in the past. In fact, the attention of science will be given to this agency of communication and travel as the many attempts in this direction to construct a practical, serviceable airship will be accomplished shortly under this influence.* From an esoteric standpoint, the Sun's progress into Aquarius strikes the keynote of unity; that wonderful symbol of the man with the pitcher of water pouring out upon humanity clearly shows that concord, harmony and unity will be the domi nane keynote of the coming country, and yet there are those of the Asiatic races who will manifest upon the lower plane of Aquarius and Urania, and there will be a division of the two, the Anglo-Saxon and Asiatic nations. Institutions of all kinds for the improvement and amelioration of the present social conditions will become more widespread, through a banding together of the more educated and thoughtful of humanity inhabiting the world in the present and coming generations, and this all for the purpose of forming the necessary environment for the coming race who are to follow; and in this way shall we begin to touch the spark of divine man and womanhood, for to be merged in humanity is to be born into divinity.

*The above was written in October, 1901, and the progress made in aerial navigation since then is plain to be seen.

CHAPTER XVIII

Directions; the Most Alluring Subject and Withal the Most Provoking in the Field of Astrological Research.

Recognizing as we must the rapidity at which this age is living, the methods now in use in the art of directing and calculating future events is certainly not perfect; far from it, and it is only too clearly recognized by the student of this science. This subject is the most alluring and withal the most provoking in the field of astrological research. At the first outset the student is confronted with a grave matter, a most crucial point in truth, and upon which so much must depend, that is, in the division of the circle of the world into proper mundane houses; and to do this there are a wonderful variety of methods, the majority of which have some rational basis difficult to be overlooked; and all of which at one time or another have been experimented upon and adopted with more or less success and accuracy, as, for instance, the systems, Ptolemy, Tycho Brahe, Firmicus, Cardan, Parphyrius, Stabilli, Regiomontames, and finally the systems of several combined in one by Zariel. These are a few only of the methods that have been in use in the art and practice of directing, and it clearly demonstrates that a great diversity of opinion has existed even in the modern practice of this science of astrology. It also goes to show the necessity for accuracy in this particular line, for it forms the base of calculating future events. Thus directing is in a somewhat muddled state at this day. There is something missing which neither an incorrect division of the houses nor the respective lordships, astral dignities, etc., of the planets can be accounted for.

It is sure that a number of methods exist and it is evident that a majority at least lead to the same final result; though it certainly becomes impossible when they do not define similar arcs in similar times, and this is what inevitably takes place when the modern student delves deep into this subject, and he becomes confused, as would only be natural. The method of setting out the twelve cusps of the mundane houses, almost universally followed by present day students of this science, is according to the method as advocated by Ptolemy and also used by Placidus and others, that is, by oblique ascension, so we may give this method some attention. We learn that the object aimed at in directing is to ascertain the arc of space that is intercepted between the degree, held at birth by a star or planet, and the decree at which, when it arrives, it will form an aspect of a particular number of degrees with another star or planet or the bodily conjunction itself, ever keeping in mind that the so-called body is circumstantially one, only regarding that it is the point of a radical position at a certain moment, and that moment is the one in which the native inspires its first breath. Now if there is any intrinsic value in an aspect it evidently is involved in its constitution of a definite number of degrees antecedent and subsequent to the forming of which power must increase in a certain definite ratio.

It follows then, that as the aspect forms it must increase in its active power in a certain definite ratio that the events that are produced by the completion of a directional arc are really the outcome of longer or lesser periods of time, and the inception being the approach of the aspect, and from this fact there will be no difficulty in comprehending the reason for such an event as the passing from the mortal expression in some instances before the arc is really completed or before the direction is finished. In this the student must be careful in rendering judgment as to the time when the active elements in the aspect begin to express themselves through their

mediums or agents, the signs of the Zodiac. It will also
be found that in close observation certain planets in certain signs and houses, and even in aspects to certain other
planets, are rendered passive in their influences or active
power, and must pass from this position into aspects with
certain aspects that will bring the thing into manifestation at a certain time that can be judged with certainty
and precision by the student, and in such instances where
the planets involved in the forming aspect are active in
their expression, then look to the significator and co-
significator, the Moon, also the promittor, and when these
are within orbs, more especially the significator and promittor, then judge that the power of the aspect must
begin to exert its influence, though keeping in mind that
the radical weakness or strength as expressed in the geniture must ever be carefully considered, so that either in
the case of one of an advanced age or with one whose
vital forces are radically low, having been so induced by
previous affliction, which will be shown in the figure,
either in the radix or progressive movements or transits,
it is then consistent to judge that the physical will hardly
remain longer a fit abiding place for the individual and
will hardly be able to suffer the strenuousness of the perfect aspect before the transition takes place. In rendering judgment upon this subject of so-called death, the
greatest care must be exercised and the student will observe how necessary it is to study the figure of birth and
the vitality as possessed by the native at the moment of
birth and whether it has been increased or lessened in
the progressive movements of the planets and stars.

In considering primary directions proper we have what
is termed both the mundane and Zodiacal arcs, though in
reality they are all mundane or formed by the diurnal
revolution of the Earth on its own axis, as will be shown
later. The distinction consists in this: that whereas the
former are measured in the world, independent of the
Zodiac, and have to do with the angles, cusps and houses

in any figure and arc taken with latitude, while the latter appear to be measured by Zodiacal degrees, the latitude of the promittor not being taken, closer distinctions will be observed as we proceed. It will be well at this point to take an exemplary nativity to the student who has been accustomed solely to secondary method with its abscence of complicated computations and semi-arcs, nocturnal and diurnal will no doubt appear somewhat confusing.

We will take an example and consider a chart* erected at a time when the twelfth degree of Scorpio is ascending. The tenth degree of Sagittarius is on the cusp of the second, the eighteenth degree of Capricorn is on the cusp of the third house, the twenty-eighth degree of Aquarius is on the cusp of the fourth house, the twenty-ninth degree of Pisces is on the cusp of the fifth house, the twenty-third degree of Aries is on the cusp of the sixth house, the twelfth degree of Taurus is on the seventh house, the tenth degree of Gemini is on the cusp of the eighth house, the eighteenth degree of Cancer is on the cusp of the ninth house, the twenty-eighth degree of Leo occupies the cusp of the tenth or mid-heaven, the twenty-ninth degree of Virgo is on the cusp of the eleventh house, the twenty-third degree of Libra is on the cusp of the twelfth house. The Sun is posited in the third degree and thirty-ninth minute of the sign Taurus in the sixth house, the Earth in the same degree and minute in sign Scorpio in the twelfth house; the Moon is posited in the thirtieth degree of the sign Virgo in the eleventh house; Mercury in the twenty-eighth degree of the sign Aries in the sixth house; Venus is posited in the twenty-ninth degree of Aries in the sixth house; Mars is in the twentieth degree of the sign Cancer in the ninth house; Jupiter is posited in the twenty-ninth degree of Aries in the sixth house; Saturn in the seventeenth degree of Sagittarius, retrograde, in the second house; Urania in the fourteenth

*See next page.

degree of sign Cancer in the eighth house; Neptune in the seventeenth degree of sign Aries in the fifth house. Latitude is fifty-two degrees, thirty minutes north; longitude is seven degrees west. The right ascension of the mid-heaven is one hundred and forty-nine degrees and fifty minutes. We find the Sun has no latitude. The declination is twelve degrees and forty-four minutes north; the Sun's right ascension is thirty-one degrees and twenty-two minutes; the Sun's mid-distance is sixty-one degrees and thirty-two minutes; the semi-arc being seventy-two degrees and fifty-four minutes. The Moon's latitude is three degrees and thirty-eight minutes north, declination, three degrees and twenty-three minutes north, right ascension is one hundred and eighty-one degrees and twenty-three minutes, the mid-distance thirty-one degrees and thirty-three minutes, the semi-arc is ninety-four degrees, twenty-five minutes. Mercury's latitude is no degrees and forty-six minutes south; declination nine degrees and twenty-four minutes north. Right ascension twenty-five degrees, twenty-eight minutes; the mid-distance fifty-five degrees and thirty-eight minutes, the semi-arc seventy-seven degrees and thirty-three minutes; the latitude of Venus is one degree and seven minutes south; declination, ten degrees and sixteen minutes north; right ascension, twenty-seven degrees and forty-seven minutes; mid-distance, fifty-seven degrees and fifty-seven minutes; the semi-arc, seventy-six degrees and twenty-two minutes. Mars' latitude is two degrees nineteen minutes north; declination, seventeen degrees and nine minutes north; right ascension, one hundred forty-two degrees and forty-five minutes; mid-distance, seven degrees and five minutes; semi-arc, one hundred thirteen degrees forty-one minutes. Jupiter's latitude, one degree one minute south; declination, ten degrees and six minutes north; right ascension, twenty-seven degrees three minutes; the mid-distance, fifty-seven degrees thirteen minutes, and the semi-arc, seventy-six degrees and thirty-five minutes.

It is the declination of the planets north or south that
is essential.   The longitude is used only in having the
given place as to time of birth.   In the examplary map or
chart as given, we find Saturn's latitude is one degree and
forty-four minutes north, the declination of Saturn is
twenty-one degrees and three minutes south.   The right
ascension is two hundred and fifty-five degrees and thirty-four minutes; the mid-distance, seventy-four ,degrees
and sixteen minutes; the semi-arc of Saturn is one hundred and twenty degrees and five minutes.

Next comes Urania, whose latitude we find to be no
degrees and twenty-five minutes north; declination of
Urania is twenty-three degrees and nine minutes north;
the right ascension we find is one hundred and five degrees and five minutes; the mid-distance is forty-four
degrees and forty-five minutes; the semi-arc of Urania is
fifty-six degrees and eleven minutes.

Then follows Neptune, whose latitude we find to be one
degree and thirty-five minutes south; the declination, five
degrees and nine minutes north; the right ascension of
Neptune is sixteen degrees and five minutes; the mid-distance, forty-six degrees and fifteen minutes; the semi-arc
is eighty-three degrees and sixteen minutes.   We have
learned that a star's mid-distance is the intercepted arc of
right ascension between the star itself and the mid-heaven
or the nadir or lower heaven and the semi-nocturnal and
diurnal arcs, the time a star takes in traveling from its
rise to mid-heaven or from its setting until it arrives at
the lower heaven or nadir.

How to Find the Mid-distance of a Star.

In order to find the mid-distance of a star, subtract the
star's right ascension from the right ascension of the mid-
heaven or cusp of tenth, or from the lower heaven nadir

or cusp of the fourth, as, for example, in the map as given, we desire to learn the mid-distance of the planet Mars. As we find this planet above the Earth, the result will be its distance from the cusp of the tenth. We have the right ascension of mid-heaven as one hundred and forty-nine degrees and fifty minutes. We observe the right ascension of Mars to be one hundred and forty-two degrees and forty-five minutes, and this number subtracted from the right ascension of mid-heaven gives seven degrees and five minutes, the mid-distance of Mars in the figure, and which is also the arc of Mars to the mundane conjunction of mid-heaven.

Then we may take another example, that is, the mid-distance of Neptune. As Neptune is below the Earth, this will be its distance from the cusp of the fourth house, being nearer to the nadir than the mid-heaven. In obtaining this result we must add the circle to Neptune's right ascension before subtracting, as it must be kept in mind that computation must ever be made in the order of the signs of the Zodiac, as follows: The right ascension of Neptune we observe is sixteen degrees and five minutes, to which is added the circle three hundred and sixty degrees, giving a total of three hundred and seventy-six degrees and five minutes, from which we subtract the right ascension of the nadir, which is three hundred and twenty-nine degrees and fifty minutes. This gives the mid-distance of Neptune as forty-six degrees and fifteen minutes.

CHAPTER XIX

Zodiacal Arcs Are Apparently Constituted by the Proper Motion of Various Stars and Planets Through the Belt of Space in Which They Move.

We will now take up a new set of arcs which may be termed Zodiacal for the reason that in place of being formed by the mundane motion of the Earth they are apparently constituted by the proper motion of various stars and planets through the belt of space in which they move. They are of various kinds, that is, the mid-heaven and ascendant direct and converse, planets to ascending and culminating degrees and the Sun and Moon direct and converse. As in the mundane group of which we have been treating it will be better to begin by bringing up all the arcs to the mid-heaven and the ascendant and then to follow them up among the various planetary bodies. In working this out the student must observe that he cannot proceed with an assumption of extracting one arc from another previously obtained as each arc will require its separate working out and in the working out of these problems the student will not be able to use the proportional parts of the semi-arcs, and though without doubt zodiacal directing will present greater initial difficulties than mundane directions, they will not be found insuperable. Beginning as in the former section, we require to direct a planet to an aspect of the mid-heaven. This is a very simple problem, that is the difference between the right ascension of the mid-heaven and the place or aspect and these taken without latitude will be the required arc. Let the student note, without latitude. In considering direct directions no matter whether the planet is on the eastern or the western side of the mid-heaven the degrees on the mid-heaven itself must advance in regular conformity.   As for instance, in directing the mid-heaven to the conjunction of the planet Urania.   In this instance when the twenty-seventh degree of the sign Cancer, that is the twenty-sixth degree and thirty-eighth minute, to be accurate, arrives upon the mid-heaven the conjunction will be formed, though, had the planet Urania been posited upon the other side of tenth cusp it could not have been so except by converse motion.

We will now take another figure* to illustrate more clearly.   The right ascension of the mid-heaven ninety-three degrees and seven minutes; the second degree and fifteenth minute of Libra is upon the ascendant; the twenty-seventh degree of Libra is on second cusp; the twenty-seventh degree of Scorpio occupies the cusp of third; two degrees forty-seven minutes upon the cusp of fourth; the ninth degree of Aquarius occupies the cusp of fifth; the ninth degree of Pisces occupies the cusp of sixth; second degree and fifteenth minute is upon the descendant; twenty-seventh degree of Aries on the cusp of eighth house; twenty-seventh degree of Taurus occupies ninth cusp; second degree forty-seventh minute is upon cusp of tenth; ninth degree of Leo on cusp of the eleventh; ninth degree of Virgo occupies the cusp of twelfth; the Moon posited in the third house in tenth degree of the intercepted sign Sagittarius; the Sun occupies the second house in fourth degree of sign Scorpio; Mercury placed in the first house in the eighteenth degree of sign Libra; Venus near in same house and sign occupies the twenty-fourth degree; Mars in the eleventh house in the twenty-ninth degree of Leo; Jupiter the ninth house in the intercepted sign Gemini, the twenty-seventh degree, retrograde; Saturn posited in the third

*See next page.

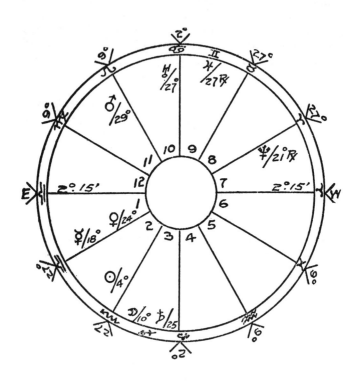

house in the twenty-fifth degree of Sagittarius; Urania placed in the tenth house occupies the twenty-seventh degree of sign Cancer; the planet Neptune occupies the twenty-first degree of sign Aries—retrogrades in the seventh house.

The Sun has no latitude. The declination is twelve degrees and forty-four minutes south; the right ascension is two hundred and eleven degrees and seventeen minutes. The mid-distance is sixty-one degrees and fifty minutes. The semi-arc is one hundred and seven degrees and six minutes. Then comes the Moon. We find the latitude of the lesser luminary to be two degrees and forty-five minutes north. The declination, eighteen degrees and fifty-seven minutes south; her right ascension two hundred and forty-seven degrees and fifty-six minutes; the mid-distance, twenty-five degrees and eleven minutes; the semi-arc, one hundred and sixteen degrees and thirty-three minutes.

Next comes the planet Mercury. The latitude, two degrees and four minutes north; declination, five degrees and three minutes south; the right ascension, one hundred and ninety-seven degrees and two minutes; the mid-distance, seventy-five degrees and five minutes; the semi-arc, ninety-six degrees and thirty-six minutes.

Then comes Venus. Her latitude we find to be one degree and nineteen minutes north; her declination, seven degrees and forty-three minutes south; her right ascension, two hundred and one degrees and fifty minutes; her mid-distance, seventy-one degrees and seventeen minutes; her semi-arc one hundred degrees and eleven minutes. Then follows Mars. His latitude, we observe, is one degree and thirty-three minutes north; his declination, thirteen degrees and nineteen minutes north; his right ascension one hundred and fifty-one degrees and thirty-seven minutes; his mid-distance, fifty-eight degrees and thirty minutes; his semi-arc is one hundred and seven degrees and thirty-one minutes.

Jupiter comes next.   His latitude we find to be no degrees and thirty-three minutes south; his declination, twenty-two degrees and fifty minutes north; his right ascension is eighty-five degrees and fifty-three minutes; his mid-distance, seven degrees and fourteen minutes; the semi-arc is one hundred and twenty-three degrees and fourteen minutes.   Saturn follows.   His latitude is no degrees, fifty-six minutes north; his declination is twenty-two degrees, twenty-six minutes south; his right ascension, two hundred sixty-four degrees, eighteen minutes; his mid-distance, eight degrees, forty-nine minutes; semi-arc, one hundred twenty-two degrees, thirty minutes. Urania follows.   The latitude is no degrees, twenty-nine minutes north; the declination, twenty-one degrees, and twenty minutes north; the right ascension is one hundred and eighteen degrees, and forty-five minutes.   The mid-distance is twenty-five degrees thirty-eight minutes; the semi-arc one hundred and twenty degrees, thirty-three minutes.   Neptune follows.   The latitude is one degree, forty-three minutes south; declination, six degrees, thirty-one minutes north; right ascension, nineteen degrees, fifty-one minutes; the mid-distance, seventy-three degrees, and sixteen minutes; the semi-arc, ninety-eight degrees, thirty-three minutes.

Examples in Directing.

First, we will direct the mid-heaven to a conjunction of the planet Urania, direct direction in Zodiac.   It will be found that the conjunction falls in the twenty-sixth degree and thirty-eighth minute of sign Cancer, this being the radical place of the planet Urania, note chart as given, the right ascension of which, without latitude, is one hundred eighteen degrees and thirty-nine minutes.   Thus we have right ascension of place of conjunction twenty-sixth degree and thirty-eighth minute equals one hundred eighteen degrees thirty-nine minutes, and subtracting the right ascension of the mid-heaven from this, we have a remainder of twenty-five degrees and thirty-two minutes, and this constitutes the arc of direction of mid-heaven conjunction to Urania in Zodiac, direct direction. It will be observed that direct motion in Zodiacal arcs implies that which follows in the natural order of the signs, while converse motion in Zodiac is really Zodiacal retrograding. In the example just considered, we find that the twenty-seventh degree of Cancer arrives in the mid-heaven by direct mundane motion, for we must realize that the mid-heaven proceeds backwards degree by degree until it reaches the radical place of Urania, though every portion of the intercepted are between such planet and the mid-heaven must have traveled over the latter ere the former really operates its conjunction aspect. We will now consider another example. We desire to obtain the arc of the mid-heaven to the square aspect of the planet Neptune, direct direction Zodiac. We find that the square falls in twenty degrees and forty-sixth minute of Cancer. The right ascension is found to be one hundred and twelve degrees and twenty-six minutes. We proceed in this wise from the right ascension of the exact point of aspect, that is twentieth degree and forty-sixth minute of Cancer. We subtract the right ascension of the mid-heaven ninety-three degrees seven minutes, and we have a remainder of nineteen degrees, which constitutes the arc of direction of the mid-heaven square to Neptune, direct direction Zodiac. In this example the student will observe that in the place of applying to the bodily impact from a position that is east from the mid-heaven, it really completes an arc of ninety degrees by mundane recession therefrom, and when the twenty-first degree of the sign Cancer comes to the mid-heaven it may be readily observed that the nineteenth degree and nineteenth minute will have been passed over the tenth house cusp or mid-heaven, or we may put it in this wise to make more clear, that is, the measure of the right ascension which is required in order to complete the square aspect of mid-heaven to planet Neptune, to which it is found to be applying at the birth. It will be observed that parallels may be considered in the same manner, that is, by taking the right ascension of the point of the Zodiac which is found to possess a similar declination as that of the planet which is evolved in the direction, computing as heretofore; though converse directions are the reverse of those as they are formed by an apparent retrograde movement in the Zodiac, or considering all arcs without distinction as being formed by the diurnal motion, and then by the opposite of this, which will be explained further.

Converse Directions Continued.

To continue our subject relative to converse directions, that is, considering all arcs without distinction as though they were formed by the diurnal motion and by such a revolution of the Earth as would necessarily present to view the Sun rising in the west and setting in the east, as for instance in the chart first given we find that Jupiter has passed the meridian, and as in these directions now being considered, it is really the mid-heaven itself and not the planet which is supposed to be the directed point, therefore as a result, it must move to the planet though in reality it does nothing of the kind. The method of working out these examples is as before in direct directions. Taking the difference between the right ascensions of the two places gives the arc, with the exception that the right ascension of the mid-heaven will ever form the minuend, and the right ascension of the place of aspect the subtrahend in place of the contrary obtaining as in the direct motion. We will now consider an example of this, that is, to obtain the arc of direction of the mid-heaven in conjunction aspect to Jupiter by converse motion in the Zodiac. We find the conjunction falls in the twenty-sixth degree and fourteenth minute of sign Gemini, the right ascension of which is eighty-five degrees and fifty-three minutes. We proceed in this wise. From the right ascension of the mid-heaven, that is, ninety-three degrees and seven minutes, subtract the right ascension of the place of conjunction, that is, twenty-sixth degree fourteenth degree of sign Gemini, that is, eighty-five degrees and fifty-three minutes. This leaves a remainder of seven degrees and fourteen minutes, and this constitutes the arc of direction of the mid-heaven in conjunction aspect to Jupiter in converse motion in Zodiac.

We will now consider another example of converse motion. We desire to know the arc of direction of the mid-heaven in sextile aspect to planet Neptune in converse motion in Zodiac. We first find in what sign, degree and minute the aspect falls. In this instance it is in the twentieth degree, forty-sixth minute of sign Gemini. The right ascension of this objective point must next be learned. This we find to be seventy-nine degrees and fifty-three minutes. Then we proceed as before, that is, from the right ascension of the mid-heaven, that is ninety-three degrees and seven minutes, take the right ascension of the place of Neptune, that is seventy-nine degrees and fifty-five minutes. This gives a remainder of thirteen degrees and twelve minutes, which constitutes the arc of direction of the mid-heaven in sextile aspect to Neptune in converse motion in the Zodiac. The student will observe that the lessening right ascension by the retrograde motion will increase the arc for the reason that it represents degrees that have passed the meridian at more or less times previous to the birth. In truth, these converse directions are positive in their effects, as a careful consideration will demonstrate, and the illustrations given should make this clear to the mind of the student. In the instance of the last example given, if we consider that the twentieth degree and forty-sixth minute of Gemini possesses, so to speak, the mid-heaven, then naturally the place of Neptune would really be in sextile aspect to this point.

We will now take up the ascendant in directing by converse motion. The student will observe that Zodiacal directions to the ascendant will take a little longer time to work out than those to the mid-heaven, as these are far more complex and involved. At the same time the opportunities are many for errors to be made in calculations; one error in the computations and the results are of no value. These directions to the ascendant are all calculated by oblique ascension in contra-distinction to the meridian ones in which is invariably applied right ascension, the arc ever being the distance between the oblique ascension of the aspect and that of the ascendant, while the latter is found by adding ninety degrees to the right ascension of the mid-heaven, and the method of obtaining the former will now be given. We will take an example, using the last map given for this purpose. We desire to know the arc of ascendant in conjunction to Venus by direct direction in Zodiac. The conjunction falls in the twenty-third degree and sixth minute of the sign Libra, and we find the right ascension of this point to be two hundred and one degrees and twenty-two minutes in proportional logarithms equals ten and one hun-fifty-nine minutes south; the ascensional differences being that between the oblique and the right ascensions. It will be observed that by adding the logarithmic tangent of the latitude of the place of birth to that of the declination of aspect, the result being the sine of ascensional difference. We proceed in this wise. The latitude of the place of birth is fifty-two degrees and twenty-eight minutes in proportional logarithms equals ten and one hundred fourteen four hundred ninety-seven hundred thousandths. Add to this the logarithmic tangent of latitude. The declination, we find, is eight degrees, fifty-nine minutes south; equals in proportional logarithms nine and one hundred ninety-eight eight hundred ninety-four hundred thousandths, gives the logarithmic sine ascensional difference of eleven degrees, fifty-three minutes, equals in proportional logarithms nine and three hundred thirteen three hundred ninety-one hundred thousandths. This, added to the right ascension of place of aspect, will give the corresponding oblique ascension. It must be kept in mind, however, that when the declination is north the ascensional difference must be subtracted, but when it is south, as in this example just considered, then it must be added. The right ascension of point of conjunction, that is twenty-three degrees and six minutes of Libra, is two hundred and one degrees and twenty-two minutes; adding the ascensional difference, which is eleven degrees and fifty-three minutes, this will give the oblique ascensions of conjunction to be two hundred thirteen degrees and fifteen minutes. We have now the element primarily sought for by the simple process of subtraction between this and the oblique ascension of the ascendant, and it becomes a simple matter to work out.

We have the oblique ascension of conjunction as two hundred thirteen degrees, fifteen minutes, and subtracting the oblique ascension of ascendant one hundred eighty-three degrees, seven minutes, we have a remainder of thirty degrees and eight minutes. This constitutes the degree and minute of the ascendant conjunction to Venus by direct direction in the Zodiac. We will now proceed with still another example and will endeavor to make this still more clear to the student. We desire to direct the ascendant to an opposition aspect of the planet Neptune by direct motion in the Zodiac. This will be a good illustration of the working of these directions. The student will observe that the opposition aspect falls in the twentieth degree and forty-sixth minute of the sign Libra, and we find the right ascension of this point in the Zodiac to be one hundred and ninety-nine degrees and twelve minutes, and the declination of the same is eight degrees and six minutes, south latitude. Then to the logarithmic tangent latitude of place of birth, which is ten and one hundred fourteen four hundred ninety-seven hundred thousandths, add the logarithmic tangent of declination of eight degrees, equals nine and one hundred fifty-three two hundred sixty-nine hundred thousandths. Then we have the logarithmic sine ascensional difference, that is, ten degrees and forty minutes equals ninety and two hundred sixty-seven hundred sixty-six hundred thousandths. Then we proceed: To the right ascension of twenty degrees, forty-six minutes of Libra, that is, one hundred ninety degrees and twelve minutes, we add the ascensional difference. Note the declination in south latitude, that is, ten degrees and forty minutes. This gives a total of two hundred nine degrees and fifty-two minutes, and this is the point of oblique ascension of opposition. Then from this point of oblique ascension of opposition aspect we subtract the oblique ascension of the ascendant, that is, one hundred eighty-three degrees and seven minutes from two hundred and nine degrees, fifty-two minutes, gives a remainder of twenty-six degrees and forty-five minutes, and this sum constitutes the arc of direction of the ascendant opposition aspect of Neptune in direct direction in the Zodiac. The student must give special attention to the quality of the declination, that is, whether it be north or south, in order that it may be known whether the ascensional difference and the right ascension are to be added or subtracted.

An error is easily committed here, and necessarily the results of the working out are valueless. The student will not find any great difficulty in working out converse directions, keeping in mind that they are ever made in the backward order of signs in the same manner as those to the mid-heaven were worked out. Through those converse directions the direct ones are calculated with the exception that in the final process there is a transposition of the subtrahend and minuend between the two oblique ascensions, that is, that of the ascendant forming the minuend as being the greatest number, as for instance we desire to learn the arc of direction of ascendant to the Zodiacal parallel of planet Neptune by converse motion. We find that the parallel falls in the thirteenth degree and twenty-sixth minute of sign Virgo.

We find the declination of this point in the Zodiac to be six degrees and thirty-one minutes north, and the right ascension is one hundred and sixty-four degrees and forty-four minutes. Then we proceed in this wise: The logarithmic tangent of latitude of the place of birth equals ten and one hundred fourteen four hundred ninety-seven hundred thousandths. To this we add the logarithmic tangent of the declination, that is, six degrees, thirty-one minutes, equals in logarithms nine and fifty-seven seven hundred eighty-one hundred thousandths. Then we have the logarithmic sine ascensional difference, that is, eight degrees and thirty-three minutes equal in proportional logarithmus nine and one hundred seventy-two two hundred seventy-eight hundred thousandths. We have the right ascension of the point where the parallel falls, thirteen degrees, twenty-six minutes of Virgo, as one hundred sixty-four degrees, forty-four minutes. From this we take the ascensional difference, that is, eight degrees thirty-three minutes. This gives the oblique ascension of parallel as one hundred fifty-six degrees and eleven minutes. From this we take the oblique ascension of ascendant and we have a remainder of twenty-six degrees and fifty-six minutes. This constitutes the arc of direction of the ascendant in parallel declination to Neptune in converse motion in the Zodiac. After going over these examples carefully, the student cannot fail to obtain the proper understanding of the working out of these directions.

CHAPTER XX

Showing Why Students Fail in Their First Attempt to Grapple with the Practice of Primary Directing.

It will now be well to give some attention to the direct directions of the luminaries, and referring also to that portion of these considerations in which was given the equivalent mundane series, the terms direct and converse hold therein a different interpretation, and without doubt it is due to the confusion which this double import of terms occasions, that the students fail in their first attempts to grapple with the practice of primary directing. The student must keep in his mind that direct motion in the Zodiac is in the natural order of the signs from Aries to Taurus, to Gemini, to Cancer, and so on through the twelve signs, while mundane direct motion is that diurnal motion resulting in an apparent track made by a star or planet from east to west. In this latter form, however, the Sun and Moon must be looked upon as being fixed; otherwise arcs are said to be converse. The direct Zodiacal arcs of the Sun and the Moon are more akin to secondary directions in a sense and in many instances in the case of the inferior planets agree with them very closely. In examining the map given, the student will observe the planet Urania occupies the twenty-seventh degree of sign Cancer and the Sun occupies the fourth degree of the sign Scorpio, and it is evident here that before the trine aspect of these two planets can be formed, the greater luminary must traverse a certain number of intervening degrees, namely thirteen, between its radical position in the fourth degree of Scorpio and the place of Urania, the twenty-seventh de-

gree of sign Cancer; therefore, when the student desires
to learn when such an aspect will come into operation
he must arrange the process accordingly, that is, so that
the result will furnish so many degrees, that is by subtracting the degree of the Sun from the degree held by
the planet Urania.

The proper method to pursue is, first of all to ascertain
the exact point of the aspect, that is, where it falls in the
Zodiac. Next learn the declination of this degree and
minute; next observe the right ascension; then the mid-
distance and finally the semi-arc. Then to work by logarithmic proportion, in this wise; as the semi-arc of the
Sun or the Moon is to the mid-distance of the planet in
question so is the semi-arc of the aspect to the secondary
distance of aspect. Then finally take the Sun or the difference of the primary or the secondary distance, which
will constitute the required arc of direction. The student
must keep in mind that if either the Sun or the Moon
crosses the upper meridian to form the aspect, then it
will be necessary to add the two distances together, otherwise they must be subtracted. For an illustration we
will take an example and direct the Sun to the trine aspect
of Urania by direct motion in the Zodiac. We first
ascertain that the aspect falls in the twenty-sixth degree
and thirty-eighth minute, to be exact, of the sign Scorpio.
The declination of this point we find to be nineteen degrees and twenty-six minutes south latitude. The right
ascension, two hundred thirty-four degrees and eighteen
minutes; the semi-arc, one hundred and seventeen degrees
and twenty-one minutes; the mid-distance, thirty-eight
degrees and forty-nine minutes. Then we proceed: as
the semi-arc of the Sun, which we learn is one hundred
and seven degrees and six minutes, the proportional
logarithm of which is nine seventy-four hundred fifty-
two ten thousandths. As this amount is to the mid-
distance of the Sun, which is sixty-one degrees and fifty
minutes, and equals in proportional logarithms forty-six

four hundred five ten thousandths, so is the semi-arc of aspect, that is, one hundred seventeen degrees and twenty-one minutes equals in proportional logarithms, eighteen five hundred seventy-nine ten thousandths, so is this to the secondary distance of aspect, that is, sixty-seven degrees and forty-five minutes, which equal forty-two four hundred thirty-six ten thousandths. Then by subtracting from the secondary distance of the aspect, that is, sixty-seven degrees forty-five minutes, the mid-distance or primary distance of the aspect, which is thirty-eight degrees, forty-nine minutes, we have a remainder of twenty-eight degrees and fifty-six minutes, which sum constitutes the arc of direction of the Sun in trine aspect to Urania by direct direction in the Zodiac. The student will observe that both the primary and the secondary distances must be taken ever from the same angle in the figure and the luminary, not having to cross the lower mid-heaven or fourth cusp in order to form the required aspect, the two distances must naturally be subtracted. Then, too, the student will observe that the first term in working proportion by the aid of logarithms must be the arithmetical complement, ascertained as previously shown, by subtracting the proportional logarithms from an integer. In obtaining the arc of direction of the Moon the method is exactly the same. In the elements necessary to be taken from the place of an aspect, as set forth at the head of the previous example, it will be found necessary to add the latitude, which must be taken into account when seeking to determine the right ascension of the objective point of the aspect to be formed in the figure, and in our next meeting will give an illustration of this so it will be clear.

In illustrating the directing of the Moon sine latitude, we will give an example by directing the Moon to the square aspect of the planet Neptune. We first learn where the square falls in the figure, and note that it is in the twentieth degree and forty-sixth minute of the sign

Capricorn. Then we observe the declination of this point, which is twenty-one degrees and forty-two minutes. Next the right ascension is obtained, which is found to be two hundred ninety-two degrees and twenty-seven minutes. Then the semi-arc follows, which is one hundred twenty-one degrees and twelve minutes. Finally the mid-distance is obtained, which is nineteen degrees and twenty minutes. Then we proceed with the problem. As the semi-arc of the Moon, which is one hundred and sixteen degrees and thirty-three minutes (the proportional logarithm of which equals nine eighty-one one hundred twenty-four ten thousandths), is to the mid-distance of the Moon, which is twenty-five degrees and eleven minutes (in proportional logarithms eighty-five four hundred sixteen ten thousandths), so the semi-arc of the point of aspect, which we have found to be one hundred twenty-one degrees and twelve minutes, equals in proportional logarithms seventeen one hundred seventy-seven ten thousandths. Likewise is this sum to the secondary distance of the point of aspect, that is, twenty-six degrees and eleven minutes equals eighty-three seven hundred seventeen ten thousandths. Thus we have the secondary distance of point of aspect given as twenty-six degrees and eleven minutes, and by adding to this distance the primary distance of point of aspect (nineteen degrees and twenty minutes), we have a total of forty-five degrees and thirty-one minutes, which sum equals the required arc of direction of the Moon in square aspect to the planet Neptune in direct direction in the Zodiac. The student will observe in this illustration that the Moon has crossed the nadir or cusp of fourth in order to complete the square aspect to Neptune. As a consequence the two distances have been totaled in order to arrive at the result of required arc of direction. In every other way the method pursued has been identical with the first example. It will also be observed that Zodiacal parallels of declination between the Sun, Moon and the planets and the aspect to the ascending or culminating degree of a geniture may be computed by a similar proceeding. It may be well to give an illustration of the latter here, as for instance, the arc of direction of the Moon to a square aspect of the degree to the ascendant. The aspect falls in second degree and thirteenth minute of the sign Capricorn. The declination of this point of aspect is twenty-three degrees and twenty-seven minutes; the right ascension, two hundred seventy-two degrees and twenty-four minutes; the semi-arc, one hundred twenty-four degrees and twenty-three minutes. The mid-distance is no degrees, forty-three minutes.

In directing the Moon to a square of the ascendant in direct direction in the Zodiac is a good illustration of working out these directions. Having learned the degree the aspect falls in, its declination, right ascension, semi-arc and mid-distance, etc., we proceed as follows: As the semi-arc of the Moon, which in proportional logarithms equals nine and eighty-one one hundred twenty-four ten thousandths, is to the mid-distance of the Moon, which equal in proportional logarithms eighty-five four hundred sixteen ten thousandths, so does the semi-arc of aspect equal in proportional logarithms sixteen fifty-one ten thousandths to the secondary distance of aspect, which is twenty-six degrees fifty-three minutes, in proportional logarithms eighty-two five hundred ninety-one ten thousandths. Thus we have the secondary distance of the aspect as twenty-six degrees fifty-three minutes, from which we must subtract in this instance the mid-distance of the point of aspect. The mid-distance we know is no degrees, forty-three minutes. After subtracting we have a remainder of twenty-six degrees and ten minutes, which sum constitutes the arc of direction of Moon in square aspect to ascendant. In this illustration it is observed that the Moon has not crossed the nadir or mid-heaven to form the aspect in question; therefore in place of adding, as in example of obtaining arc of direction of Moon in square to Neptune, the mid-distance must be taken from the secondary distance, while in the former example the primary distance is added. After a careful study of these illustrations the student should be able to solve any problem coming under this head. We will now give some consideration to the converse directions of the two luminaries. As in other converse directions, of which we have treated heretofore, these are made backwards in the Zodiac in reality, though of themselves alone these directions are not of great importance. However, when coupled with other aspects of similar influence and effect in the horoscope, they must be reckoned with. The method is the reverse of direct directing, as the same elements are obtained in the first instance though the formula for the rule of three to be performed is, as the semi-arc of the aspect in question is to the mid-distance of the aspect, so is the semi-arc of the Sun or Moon, whichever it may be, to the secondary distance of the Sun or Moon. The next procedure is to either add or subtract as in the previous illustration given, that is, the two distances of whichever luminary is involved in the example. We will now consider an example of this, taking the figure given last. Desired to obtain the arc of direction of the Sun to a conjunction aspect of Venus by converse direction in the Zodiac. We first learn the point in which conjunction falls, that is the twenty-third degree and sixth minute of sign Libra; the declination is eight degrees and fifty-nine minutes; the right ascension two hundred and one degrees, twenty-one minutes; the semi-arc is one hundred and one degrees, fifty-three minutes; the mid-distance is seventy degrees, forty-five minutes. We then proceed as follows: As the semi-arc of aspect, that is one hundred one degrees, fifty-three minutes, which equal in proportional logarithms nine and seventy-five two hundred eighty-three ten thousandths, as this sum is to the mid-distance of aspect which in proportional logarithms equal thirty-nine nine hundred forty-five ten thousandths, so is the semi-arc of the Sun equal in proportional logarithms twenty-two five hundred forty-eight ten thousandths.

In considering the various branches of this science, we have the same basis for all though the influences vary in their effect and as a result. It is possible to treat upon each separate and apart from one another, except as in the horary branch, which is very closely allied to the natal branch.

It will be possible to consider each branch of the mundane as well as astrometrology. All are important and a thorough knowledge of the first and most important, the natal, will remove the difficulties to a clear comprehension of all other branches. The semi-arc of the Sun as given to the secondary distance of the Sun, seventy-five degrees and twenty-six minutes, equals in proportional logarithms thirty-seven seven hundred seventy-six ten thousandths. Thus we have the secondary distance of the Sun as seventy-five degrees twenty-six minutes from which is subtracted the mid-distance of the Sun, the mid-distance being sixty-one degrees and fifty minutes. This leaves a remainder of thirteen degrees and thirty-six minutes, which is the arc of direction of the Sun in conjunction aspect to Venus in converse motion in the Zodiac. It will be observed that the converse directions of the Moon are exactly similar and proceed along the same lines, as also do the converse arc of planets to the degree or ascendant and mid-heaven. However, we will consider an example here of this: Desired to obtain the arc of direction of the ascendant in square to planet Urania by converse direction in the Zodiac. We find first that the aspect falls in the second degree and thirteenth minute of the sign Cancer, the declination of which is twenty-three degrees and twenty-seven minutes, the right ascension of which is ninety-two degrees and twenty-five minutes; the semi-arc is one hundred twenty-four degrees and twenty-three minutes, the mid-distance is no degrees and forty-two minutes. Then we proceed as before: as the semi-arc of the point of aspect, which in proportional logarithms equals nine and eighty-three nine hundred forty-nine ten thousandths, is to the mid-distance of point of aspect, equal in proportional logarithms to two and forty-one seventeen ten thousandths, so likewise is the semi-arc of Urania, equal in proportional logarithms to seventeen four hundred eleven ten thousandths, to the secondary distance of planet, Urania, which is zero degrees, forty-one minutes, equal in proportional logarithms to two hundred forty-two three hundred seventy-seven hundred thousandths. Thus we proceed. Then the mid-distance of the planet Urania, which is twenty-five degrees and thirty-eight minutes, must be added to its own secondary distance, which is no degrees, forty-one minutes, the total of which is twenty-six degrees and nineteen minutes. This sum constitutes the arc of direction of the ascendant in square aspect to planet Urania by converse motion in Zodiac. One of the essential points to keep in mind is to be sure to take the proper semi-arc of the directed planet, as there may be a change from the nocturnal to the diurnal arc in computing the aspect by reason of the passage of a body to a position above the horizon, or the change may be the reverse as from a point in elevation to a station below as given, intimating also one from the diurnal semi-arc to the nocturnal semi-arc. We have now covered quite closely, at least, the mathematical portion of the primary directions. The student will observe that as Zodiacal and mundane directions are made in two different circles and by two different motions, the Sun and Moon will meet all bodies that possess latitude twice, first by motion forward through the ecliptic (all star rays being encountered in that pathway with latitude), whereas in the latter the significator is supposed to remain fixed in its hour circle and meet rays by the diurnal motion of the Earth.

The prorogator should never be directed except to the descendant or western angle, as the prorogatory place is the point of the horary circle in which the prorogator is found at the moment of birth and which point is ever radical, that is, it ever retains the same relative position to the meridian and also to the equator. Therefore the student will realize that it is consequently immovable and is exposed to the bodies or rays of the promittors, no matter whether they be good or so-called evil in nature, as they arrive there by mundane converse motion, and for this reason the latter form of directing is so powerful in its effects, though the student must ever keep in mind that it is the mundane circle and not the Zodiacal one that is referred to; for in the latter, as stated heretofore, direct motion in the Zodiacal series of directions is by far the most important and powerful in its effects. Both the mundane and Zodiacal aspects were considered somewhat by ancient students of this science, and the mundane motions were rediscovered by Placidus. However, in working out of these directions there was much controversy at that time, each one intent only upon his own pet theory. In fact, all held some parts of the truth and had they combined their efforts and exchanged views they could no doubt have arrived at the real truth and by casting off some unnecessary details could have utilized both the mundane and Zodiacal motions and directions. To be sure, the applying and separation of stars and planets is largely mundane. The student will observe that certain points in the figure ever hold the same aspects to one another, as for instance the ascendant and mid-heaven are ever in square aspect; the descendant and ascendant are ever in opposition aspect, while the ascendant and ninth house are ever in trine aspect. The same is true of the ascendant and the fifth house, the fourth house or nadir being in square, as is the mid-heaven, while the eleventh house cusp, also the third, are in sextile aspect.

Then, too, these configurations hold in other points in the figure.

Rapt Parallels Not Understood Since Chaldeans' Time.

Relative to mundane and rapt parallels, these were not understood since Chaldean's time. The student will observe that by mundane motion each temporal house obtains its value entirely from a proportional part of a star's semi-arc, which must ever be a third, either nocturnal or diurnal, and the same holds in spherical trigonometry each section of a sphere made by a plane is a circle and it follows that these mundane houses are constituted by a series of great circles above the planes which cut the sphere of the heavens and form the boundaries of the various houses. It is imperative that the student should thoroughly understand this, as well as other facts associated with the sphere, as such knowledge will tend to simplify the matter of primary directing. A knowledge of spherical trigonometry is necessary, mundane motion being caused by the diurnal rotation of the Earth from west to east, which is completed in twenty-four hours, and which is responsible for the alternating effects of day and night, it naturally follows that it is uniform; therefore when one direction is given, others may be easily obtained and deduced by a proper use of the aliquot parts of the semi-arc. On the other hand, the student will observe that in the Zodiacal motion the appulse may be interrupted by a retrogression, stationary attitudes, slow movement, or it may be quickened by the adoption of a higher rate of progress than the mean.

Placidus was able to find expression for material conditions. In giving this data, you must realize that we have the advantage of conferring with individuals who have attained wisdom through the careful study of this

science and covering a long period of time. As I have
stated heretofore, it is impossible to obtain accurate
knowledge of the influences of all planets in the ordinary mortal life of three score and ten years. Considering Zodiacal motion, the student must realize that taking
all these into consideration, Zodiacal ares obtained in
the usual manner appear artificial as contrasted with
the natural and regular motion which is manifested
in the mundane motions, and the student must realize
sooner or later that all the primary arcs are mundane
and determined by the motion of the Earth. Thus, for
instance, when one of the luminaries is found above the
eastern horizon and another planet just below it, then
in directing that luminary to a conjunction of the required planet in the Zodiac, it will ever be necessary to
take the diurnal semi-arc of the place of the required
planet and this clearly demonstrates the fact that all directions are formed by the rotation of the Earth. Thus
in truth, instead of the luminary descending into the first
house, as it would necessarily have to do in order to form
the conjunction of planet there by Zodiacal motion, the
planet itself would ascend above the eastern horizon to
the place of the luminary, and the same would be true
of any body, star or planet, or of any particular point in
the figure. Therefore, having established this fact, the
student will observe that the only purely Zodiacal arcs of
directions are those formed by secondary motion after
birth and that in reality all primary arcs are mundane,
inasmuch as when computing direct directions in Zodiac it is not in reality the luminary which is directed,
but the place of the aspect is the point to be considered
in the calculation. Thus when directing the lesser luminary it should ever be considered without latitude and
not otherwise, as is taught in the present time by students of this science. It may also be observed by the
student that angles cannot be conversely directed, as
they only receive the rays in the Earth but not parallels

or rays in the Zodiac, also that the other significators, by a direct motion, receive the rays and parallels both in the Zodiac, also in the Earth, though by a converse motion the rays only and the parallels in the Earth but not in the Zodiac. However, it must not be concluded from this that there is no virtue in the directions to angles, and the Zodiacal directions to angles must be considered, though many of the more modern students deny this and, no matter whether mundane or Zodiacal, direct or converse, including the parallels of various kinds and natures. Thus the student must not cast out any one essential part but hold fast to that which promotes accuracy, no matter whether it be considered of minor importance. Every part is essential to the whole, and it is only when all are taken together as one and judged as such can accurate results be found. The student must first of all be unbiased, and not held by any hard and fast lines that cannot be demonstrated as truth when used in a practical manner.

CHAPTER XXI

Positive Secondary Directions Are Quite Powerful Factors in Producing Events and the Student Will do Well to Look Upon Positive Primary Directions in the Same Manner.

It must be realized that converse directions to angles are formed even before the birth, as we have stated. Positive secondary directions are quite powerful factors in producing events, and the student will do well to look upon the positive primary directions in the same manner as the effects of their influences will show that they, too, are of vast importance and must ever be given careful consideration. Then, too, another point of importance must be regarded in computing parallels, and that is that their effects will generally precede their formation. This is owing to the magnitude of the Sun and Moon, by which their bodies are effected before they are completed, center to center. This may not always operate in every case, depending much upon position of bodies and the aspects that excite into activity, and the results of these can be attributed more to aspects than to parallels, and in some instances a small equation might be utilized in order to render more accurate the exact time of aspect, that is, when it has attained to greatest intensity. In the case of birth of children that do not survive infancy, the student may often find instances where there is no apparent manifestation of aspects from secondary directions, though often a configuration may take place and the student will observe in such instances that the most satisfactory and effective aspects in such instances of infantile mortality, are caused from primary arcs of direction. In many writings upon this particular subject the student will observe that the results are due to position, as for instance, the child may pass from the mortal by position, meaning that the child at its birth, as set forth in the natal chart, may have a conjunction, square or opposition of the luminaries, one or both to Saturn. The student, however, must not take this to mean that all must hold the same degree and minute in the same sign of Zodiac, as in signs of opposite or square aspect, for in truth they may only be within orbs or from three to five degrees, and then by taking one degree for each year it would follow that the aspect would not be complete until the space had been closed between the planets or bodies in question, though the student will observe that in the meantime, other things concurring and conducing thereto, the native might pass from the mortal, from there being no perspicuous operative direction complete at the time. However, the principal aspect was forming and only required an influence of similar nature to bring it into action. The anaretic arc is a gradual planetary appulse from platic to partile conjunction or aspect of power, the anaretic arc being the distance that the two bodies stand originally from one another, which equated carefully should point out the exact period in which the aspect would be in power and take the life from the physical. However, it may be possible that the native may have been born into physical life with such a low vitality that even before the full power of the major arc of direction is completed by the conjunction or aspect in force, so-called, death may take place, more especially if other directions concur or assist in this in any way whatever. As we have stated heretofore, the student must learn this fact by careful observation of the strength of the first house, the ruling planet, also the ruler of the tenth house, its aspects, the strength of the luminaries, more especially the Sun, and the configurations directed to the eighth house.

In a large majority of instances these early arcs will manifest their power and influence very closely to time and will often receive the most startling results, such as would tend to convince the beginner to believe in a fate that was inexorable and uncompromising. It will be well at this time to make mention of the hyleg. The student will observe, in determining this point in the figure termed the hyleg, an evident confusion and ignorance on the part of many writers, it being confused with the part of fortune, the dragon's head and tail, the ascendant, mid-heaven, dispositors, etc., and there remains so much of an artificiality associated with it that it is only natural that many students should totally ignore the rules laid down by various students who have written offhand in their selections of the Aphita or giver of life, and prefer to follow the hermetic views in relation to this subject. In relation to the part of fortune there is no question of its influence to a certain degree in the figure. However, it is an error to associate this point in the figure as associated with the hyleg. True, it might have some effect indirectly, but should not be considered as hyleg. The luminaries, that is the Sun and Moon and the ascendant or first house, have to do with the hyleg. Naturally the ascendant would include the sign and planets thereon, and in aspect thereto. The Sun and Moon are considered when one or both occupy one of the hylegical places, and these are as we have given elsewhere, all of the first house, also the tenth, the fourth and eighth houses, and that half of the eleventh house which is nearest to the tenth or mid-heaven, though in each instance an orb of five degrees must be considered and must be computed by oblique ascension, so that the first house limits would be from ten degrees above the cusp of ascendant to thirty degrees below it, and the same is true of the other houses mentioned. Those houses not mentioned cannot be occupied by the hyleg, that is, ordinarily; therefore the luminary which possesses any portion of these divisions of the horoscopal figure is accounted the true prorogator and as such is directed. In such instance, where both the Sun and Moon are found to occupy hylegical places, either the Sun or the luminary aphetically strongest by aspect is chosen, and if neither the Sun nor the Moon holds such positions then the office falls to the ascendant. Now, as the ascendant cannot meet aspects in the Zodiac and as Placidus further states that even mundane aspects (this does not refer to parallels) can only be received by direct motion, it frequently occurs that under such restrictions few directions can be formed by the ascendant, and if it should happen to be hyleg an ordinary lifetime might not provide affliction sufficient to cause or account for the passing out from the mortal. However, these conditions and influences do really take effect, and it is useless to ignore them. An experience will clearly demonstrate to the student that there is much verity in the apheta chosen according to the foregoing rules, although the student may find it a little difficult to follow in the beginning. Continued practice will clearly demonstrate its efficacy, and we must consider electricity to be emanations from the photosphere of the Sun and life from the lowliest to the most complex to proceed from thence and thence alone. It signifies naught in respect to the sex in which it happens to reside, for the real life principle is never sexed.

Relative to sex of life principle, we have learned that the Moon, owing to dominion of certain functions, exerts a powerful influence over the physical expression of the female organization, though we must necessarily recognize the power of the great solar orb, the Sun, from which those radiations of life forces emanate in harmony with the expression of each star and planet body as it comes into certain angles of position, and so we must concede this great luminary as the giver of life. At this time I desire to call the attention to another term of some importance in our further pursuit of this subject of directions, that is, the parallax. This term is applied to a star planet or body when wishing to describe the angle between the directions of two lines drawn to it, that is, the one drawn from the point of view of the observer while the other is drawn from the center of the Earth planet to the star in question, as for instance, draw a circle* which will represent the Earth, then in the center place the letter C representing the center; then from the center draw a straight line to the exact center of top of circle and similar to point of tenth cusp in a horoscopal figure, and this point will represent the point of view of the observer. Now the observer or student is looking upward in space to the right or left as the case may be at some planet or star in space. Now draw a small circle to represent this planet, and then draw a line from the point of the observer or cusp of tenth direct to the circle or planet. Then again, from the planet draw another straight line to the center of the circle representing the Earth planet. Now we have three points, first the center of Earth represented by letter C, then we have the center of Earth at top represented by S, the student, then we have the small circle in space represented by P. Then the angle SPC is the parallax for the student at the point marked S. It will be quite evident to the student that the larger the body from which another is viewed and the nearer the body under inspection, the greater will be the parallactic angle, while on the other hand the smaller the globe from which the observation is made and the farther the distance of the orb in question, the more minute and inappreciable it will be. Then it naturally follows that it will be at its maximum when a star or planet is on the horizon and absolutely nil when in the mid-heaven or directly over the center of the Earth and the point of observation of the student. Thus it is seen that different mundane posi-

---

*See next page.

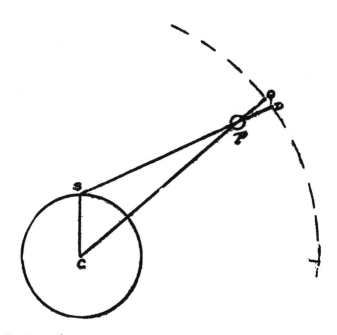

C - Earth's centre

S Place on Earth's Surface.

P - Planet or Star.

Q - Planet or Star's Geocentric long

D - „ „ Observed long.

tions will cause differences in the amount of parallax which would necessitate varying factors in the calculation of any heavenly body unless some definite point should be decided upon, to which all others might be referred, and such is the case, for the student will observe that astronomical positions are ever calculated to the center of this Earth planet. Then, too, if the native at the moment of birth into physical expression is considered to be in the center of all extra mundane force, and if aspects are to be calculated to the center of this Earth planet, there must necessarily be an equation of these positions, and at this point we come to another question, that is, the exact division of the mundane circle into the twelve houses. To do this it is quite essential to know from which point they must take their rise or ascent, that is, from the observer's superficial and tentative position, or from the mathematical center of this Earth planet.

We may realize that the child or native is for all purposes the focus of planetary influences in so far as the horoscopal figure is concerned, and this no matter whether he or she be situated at the poles, the equator or elsewhere, and as the native cannot dispose himself centrically he need not bother about such center which is evidently for him. The student must recognize then two horizons, the plane of the one passing through the eye of the observer and may be termed the sensible horizon, while the other passes through the center of the Earth planet and may be termed the rational horizon; therefore, as the diameter of this Earth planet is about eight thousand miles it naturally follows that these two planes cut the sky at a distance of half that distance apart and these are extended illimitably into space, since they are parallel. However, on the infinitely distant surface of the celestial sphere the two traces sensibly coalesce into one great single circle, which is the horizon as first described. Therefore, in fact, while man can distinguish between the two horizontal planes, we get back to one horizon circle in space, and while this may not be exact theoretically, practically it is partially true, and in some cases entirely true, for all ordinary purposes. Practically the Earth may be considered as a point in the great ocean of space, and though all influence is exerted to the central point, the fractional differences introduced by the superficial observer's point of view are too small to cause any great error in computation. We may now take an illustration that will make these points more clear to the mind of the student.

First,* draw a large circle similar to the outer circle of a horoscopal figure. Then in the exact center of this circle draw a small circle to represent the Earth, the poles outward. Then quarter the circle; at the ascendant place the letter A, the descendant, the letter D, at the tenth cusp or Zenith the letter Z, at the fourth cusp or nadir, the letter N, in the center of the small circle the letter G, at the bottom of the small circle representing the Earth planet the letter M, at the top of small circle the letter S. This is the point of view of the observer. Next draw a line from left to right of the large circle which just touches the top of small circle and point of view of student. This is what we may term the sensible horizon. At the end of this line, at ascendant, put second A and at right end a second D. Where the ascendant line proper intersects with small circle place the letter L, and where the descendant intersects with small circle in center place the letter C. Now, beginning at eleventh cusp, draw line downward at right angles making it intersect with point of view of student. The same may be done beginning at the twelfth cusp, also at the ninth and eighth cusps except they be drawn to left angle, making all these lines intersect at point of view of the student and continuing onward to their contact with large circle.

---

*See next page.

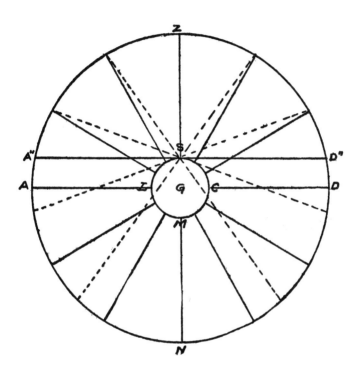

This figure, when drawn properly and completed, will show to the student at a glance the two varied positions of the mundane circle into houses. It will be observed that the division has not been made from the center with the exception of the four cardinal points. This is for the purpose of eliminating, as far as possible, the element of confusion which would be created by so many lines.

In considering the illustration as given, it will be seen that the letter S is the point of view of the student. The second A and B is the sensible horizon, the first A and B representing the line running directly through the center of small circle or Earth is the rational horizon, and that the Earth is rotating in its usual manner from west to east or right to left at or about the time of the equinox. At the point indicated by the letter L, it is noon or 90 degrees west longitude, and it is understood to be 6 A. M. 90 degrees east longitude is the same, or 6 P. M., and 180 degrees either east or west, it is midnight; 45 degrees west, it is 9 A. M. Thus, before the student at S can catch sight of the orb it must rise above his horizon, that is, the line indicated by the second A and B, disregarding refraction for the purpose of illustrating, though it does not do this in reality. As the student will observe what really takes place is this, the horizon indicated by second A, and which we may term second horizon, becomes deflected or extended in the direction of points marked by letter S and first letter A, although the Sun appears to be just elevating itself above the point marked second A, therefore naturally a small discrepancy takes place, not of course as regards the opposite point but those points that are involved in the quadrature, for the reason that it cannot be exactly 6 o'clock in the morning at point indicated by the letter S when it is hidden altogether at point marked L, though this very small amount of disagreement is by far too small to cause much variance in the actual effects so far as the calculations of the student are concerned, as the Sun really has not possessed any deducible parallax for a long time by the usual methods, at different periods of time. Various students of astronomy and astrology have given their results, as I understand, used at this time by students in your land, is eight and eighty-five tenths seconds; however, the correct figures and that used by students of ancient times, is eight and eighty-three tenths seconds. The student must ever keep in mind that the horizon as indicated by the first A and B, the rational one as regards the point of observation marked S, is not a horizon at all for an individual stationed at point indicated by letter L, though it is in reality his zenith. In this case, then, the points indicated by letters Z and N would become the rational horizon, and a parallel plane to these, or a line running up and down to outside large circle and cutting through the point of view at student at L, would become the sensible horizon. It is quite essential that the student should comprehend this fully, as it will assist him to a great degree in grasping the idea of angle of influences, etc. If it be assumed that the astronomical coincidence of the two horizons named at vast distances, there is no need of trouble over the problem whatever, though it will appear clear to the student, after careful consideration, that if it be admitted that electricity comes as light in straight lines to the Earth, and for ilustration a star or planet be posited at point indicated by letter Z, which would be removed from the Sun at first A, a distance of 90 degrees; then the square aspect will be formed at point marked by letter G, or the exact center of the Earth, and not at the point as indicated by S, the point from which the observation is being made. This is one of the essential points that I desire the student to consider well.

Rays of light, when reaching the Earth's atmosphre, are in a measure turned from their true course, or, as we may term it, refraction takes place, though this is really only a visual matter and there is no reason to suppose that the rays of influence as cast by a star or planet should

be disturbed to any great degree through any such contact or should undergo any change of direction in passing through the atmosphere.   The student must keep in
mind that there is a constant emitting of influence from
all planets to the Earth both day and night, be it for so-
called evil or good.   As the Earth is so small a body in
comparison to the great field of space as contained in
this solar system, the focalization of the rays of influence
would really involve the entire planet from a mundane
point of view, though as these rays of influence must necessarily be transmitted through the twelve Zodiacal signs.
It follows that there must be subdivisions of these rays of
influence, depending upon that sign through which the
ray was manifesting itself to whatever location or place
under the rule of that particular sign of the Zodiac.
Therefore as the student, at point of view in the illustration marked S, would see the Earth being relatively so
small a body in its relation to the great bodies in space
across which light and influence comes, it really becomes
little else than a point at the apex of a triangle, and certainly this relationship is true as relates to the majority
of the many fixed stars which are even too remote to
manifest a parallax, and the student will observe that it
is only when he considers the planetary region that the
parallax of the planets becomes manifest.   The mean parallax of the lesser luminary, the Moon, is as great as
fifty-two minutes and two seconds, while the horizontal
parallax of the planet Mars is only about twenty-four and
sixty-four tenths seconds.   Thus the student will observe
the difference caused by variance in distance of various
planets and stars.   Thus the main points to be carefully
considered in this important matter of commentation are
that it increases the right and oblique ascension and diminishes the equivalent descensions, and that it diminishes
the northern declination and latitude in the eastern part
of the figure, and increases the declination and latitude
in the western part, and increases the southern in the

eastern and western part, and diminishes the longitude in the western, and increases it in the eastern portion of the figure. Thus its effects are just the opposite to those of refraction. For the sake of perspicacity, we may consider an illustration of this, as for instance some altitude above the horizon in a given latitude, keeping in mind what has been stated relative to the parallax being greatest at the horizon and its minimum at the mid-heaven, where it only affects the declination.

For example, we desire to know the Moon's parallax when its declination is twenty degrees and twenty minutes north and the latitude is fifty-two degrees and twenty-eight minutes north, while the horizontal parallax being at this time fifty-six minutes. The student must first of all find the altitude of the equator, which will be the complementary number and so form with latitude ninety degrees.

First we take one-quarter of the circle, or ninety degrees, and from this amount subtract the latitude of the Moon as given, which is fifty-two degrees and twenty-eight minutes north. This gives a remainder of thirty-seven degrees and thirty-two minutes, and this equals the height of the equator. Then as the Moon is in north declination, it will then be above that circle, so that the declination must then be added to the height in order to obtain the correct altitude of the Moon. Then we proceed in this wise. The Moon's declination is twenty degrees and twenty minutes north, to which amount is added the height of the equator as previously obtained, or thirty-seven degrees and thirty-two minutes. We have a total of fifty-seven degrees and fifty-two minutes, which sum equals the Moon's altitude. The next procedure will be to add the logarithmic sine of the horizontal parallax of the Moon and the logarithmic cosine of the Moon's altitude together, and the total sum will represent the logarithmic sine of the parallax corresponding to that altitude. Thus the logarithmic sine of horizontal parallax

of Moon, that is, fifty-six minutes, equal eight and two
hundred and eleven eight hundred ninety-five hundred
thousandths plus the logarithmic cosine of altitude of
Moon, fifty-seven degrees and fifty-two minutes, equal
nine and seven hundred twenty-five eight hundred twenty-
three hundred thousandths. This gives seven and nine
hundred thirty-seven seven hundred eighteen hundred
thousandths, which sum represents the required parallax
of the Moon of thirty minutes. All the other elements
must be considered in a similar manner. The parallax will
naturally make some slight variance in computation if it
be considered that each and every individual be the
center of the planetary influence, though whether it would
be of great importance the student can readily judge.
He must realize that accuracy is of the greatest importance in obtaining the exact ascendant, which is not so
difficult when the exact moment of birth is known. After
the student has obtained the arc of direction there still
remains a point for consideration, that is, in the equation
and timing thereof. As it stands, it is in degrees and
minutes of space, and the student must realize that these
must represent an equivalent in time. The method
formally adopted a few centuries ago was to ascertain in
what number of equatorial degrees the succeeding place,
that is, whether it be the body or aspect of a planet will
arrive at the preceding place in the figure. As these degrees pass the horizon, as well as the meridian, all distances must be calculated according to the various positions of the planets in question with respect to those
angles. Each degree signifies one year of time. This
method has the feature of extreme simplicity, though it
will not always prove absolutely correct. However, if
the exact moment of the birth is in doubt or unknown,
this method would no doubt prove sufficiently correct for
all purposes, thus allowing one year of life for every degree of arc. In this wise, arc of direction twenty degrees
fifteen minutes, would equal twenty years, three months

no weeks.   Nineteen degrees forty-five minutes equal nineteen years, nine months, no weeks; fifty-three degrees three and one-half minutes equal fifty-three years, no months, two weeks; four degrees thirty-three minutes equal four years, six months, two weeks, the minor periods being in proportion, that is, five minutes of arc of direction, that is, one-twelfth of the whole degree, would be equivalent to one month of actual time, and so on, as examples given.

CHAPTER XXII

It is the Quality of the Thought, the Moral Power that Renders the Magnetic Forces Firm and Potent for Good of Self and Others, so that He is Proof Against the Evil Influences that May Prevail About Him.

It is true, as we have learned, that all individuals, spirit as well as mortal, are constantly creating an influence, atmosphere, aura, or we may say a sphere of life, about themselves that may be of the most refined quality of spiritual power and composed of all the elements of good. Then, again, it may be tinged with the lower qualities of undeveloped good that mars the pure life expression to that degree that the poor soul seems swallowed up in the dark clouds of despair. This atmosphere or sphere of life acts as the agent or medium through which the external world vibrates and pierces to the soul at times, the manifestations of that lower expression.

When this atmosphere is pure and clear, peace, love and wisdom prevail; an impure atmosphere, unrest, weakness and ignorance prevails. When this external, vital atmosphere is positive and is thereby superior to all external influences of a lower order, the individual not only possesses a high degree of physical health, but the mind is in possession of the most elevated order of thought. Then, too, it is the quality of the thought, the moral power that renders the magnetic forces firm and potent for good to self and others, and he is proof against the evil influences that may prevail about him. In many instances it is thought that intelligences disembodied are continually seeking to injure mortals, but in such instances it will be found that the mortal thus assailed or influenced has himself created that atmosphere that permits such influences to enter, in fact, invites such through the every-day act and thought.

Thus it is in the sphere of the individual himself that we find impurity and imperfection, an excess of the animal over the spiritual, otherwise a contact with the lower spiritual intelligences would have benefitted the intelligence disembodied rather than injured the mortal. Thus the so-called victims of obsession, as it is termed, are suffering from a self-induced organism, causing the mind to dip into moods prone to brood until the dark finger of morbidity is manifesting in the atmosphere, and in case such an individual seeks to develop mediumship, a silly, morbid wonder will generally become the keynote and emotion, which in the superior temperaments is the prelude to active wise expression, and comes to be valued for its own sake, that it entirely and completely quenches the practical impulses it should have vitalized.

It is ever found that in all normal balanced natures emotion is not separated from action, but manifests in perfect harmony and unison. Those poor souls who are the subjects of influences which only bring unhappiness and lower expressions of thought, should be taken in hand. These unhappy victims, we may say, of unhappy subjective sensations should be thrown into the companionship of those possessing cheerful and practical, as well as pure radiations, where they may be fed whenever their susceptibilities offer an opening, and should be reared in surroundings and environments where energetic, active interests prevail.

The question may be asked, what real harm can the continued presence in the impure atmosphere of a spirit of low order bring about? The result is that irrational or fragmentary thoughts are forced into consciousness independently of the will, and that irresistibly, and cause much distress, though it is an error to charge every un-developed thought to disembodied spirit.

Verbal obsessions are those in which the thought comes under the afflicted Mercury-Saturn-Mars influence. Ob-scene language, anger and possibly violence is the result. Inciting obsessions apply more to this latter expression. In such case the natural brain power is weak and un-balanced, that is the region of the brain giving impulse is vastly stronger than the moral and intellectual power, for the very reason that it has been more developed. The individual has been allowed to grow wild, as we would say of some specimen of beautiful plant that had grown imperfectly, smothered by the weeds of ignorance.

This subject cannot be considered lightly, in fact a careful study of this, and active efforts to nourish and bring to perfection such intelligences, will serve to bring peace on Earth and good will towards all more quickly than any other expression from mortal.

It is most essential to get the mind of the individual to manifest on a higher plane of expression when the state termed obsession is present, otherwise the poor soul is preparing to enter the higher life in an earth-bound state. The ordinary preaching will not change him. A practical knowledge of the higher life will assist him, but in the general conditions of individuals who are manifesting on a low plane, it is an utter impossibility for them to enter an elevated spiritual sphere of life. The true student of this divine science of life, will possess ideal health in mind and body. He will not leave his children in the care of dogmatists and permit them to be taught the au-thorized false teachings, but will instruct the young mind in the knowledge of self, physically and mentally, and make the evolution of health and well-being his chief study, in the place of a concern for safety in the life after so-called death.

There must be a greater effort to protect the young mind from developing morbid fancies that lead to a disorganized being. Character should be the objective point; spirituality of mind and a pure personal sphere and selfishness can never go together. It is the duty of all to purge the mind of silly superstitions, evil influences, reimbodiments, and attain to moral earnestness, for the regeneration of all the motives and forces of life, bringing humanity face to face with the sublimities of the natural and spiritual universe.

In the study and application of the science of Astrology it is far from satisfactory to know a certain thing will occur and not be able to figure out the time of event as well as the manner in which it will overtake the native, as when the approximate time is known, whether the event may be intimated or not, reparations may be made, and yet, so far as the operation of the heavenly force is concerned, there is no question but that the individual may grapple with it as successfully as did Eurythion, Girgon, Theseus and others.

At the present time it is impossible for the student of this divine science to understand just how the opposing extra mundane power acts, nor the point of its incidence, or even the energy required for its successful resistance. That the physical factor is not concerned so much as the psychical one, even in the production of numerous experiences that appear to act solely upon, and in fact to proceed from the rational plane, is beyond question, and innumerable occasions which work out on the physical plane have their efficient causes deeply rooted in the psychical side of man.

If the Astronomer Makes an Error, He is Pardoned,
For It is not Such a Vast Incomprehensible Study,
but the Student of Astrology Must Give Perfection
Only.

Many students are just beginning to realize this, and
the key is held in the wisdom of the divine science of
Astrology. Directing the planets in the horoscopal figure
is by far the most difficult portion of this divine science,
not alone as relating to the element of time but to the
nature of the event produced by the arc itself, and herein
the student finds his work arranged for him. The ordinary individual, who never thinks seriously along these
lines, will have none of it unless the mark of infallibility
be set thereon, not even though the seal of this remarkable
quality be thoroughly satisfactory. If the astronomer
makes an error he is pardoned, for is it not such a vast,
incomprehensible study; but the student of Astrology
must give perfection only. Both are useful, the one deals
with clay, the other with the soul; one is the anatomist,
the other the physiologist; the one scans the bars of the
scheme, admires their regularity and efficiency, the other
peers beyond into their purpose of being.

It is considered that so far as regards the necessary
calculations to be made in order to deduce the period of
operation of certain magnetic or astral forces or waves,
the student will be at fault, for if there is error made in
this point it cannot be expected that stars and planets
should hold with them. However, separate and apart
from this, if the birth moment has been accurately taken,
the horoscope correctly cast and the directions perfectly
computed, the exact time of the formation of planetary
conjunction or aspect should see the event in full manifestation, and just what that event was to be was expressed at the moment the native first drew into its lungs
the astral breath and enshrouded in secret in its sympa-
thetic aura or vital atomsphere awaiting opportunities for

expression, which are ruled solely by planetary rays and the manifestation of the free will, while the problem of discovering and naming lies solely with the astral student, that is before it comes into full or partial activity, as afterwards the native is often made only too acutely aware of its presence, its nature and its parting sting.

Therefore, as it is given into the hands of man the *modus operandi* of regulating the physical expression of an individual according to these periods, there remains alone the greatest of all, the art of warding or receiving the so-called evil or the good influence until this knowledge is correctly understood, so that it can be worked out in a practical manner, the reign of peace will not come.

We learn that the two most pregnant aids are the conquering of the external nature, or rather the internal, as directed toward externalities and knowledge; that is, knowledge that comes esoterically and is concerned not with the transitory form changing brain mind, so much as with the human soul. If man could but grapple with that long expressed truism, what ye sow that shall ye also reap, and comprehend it in its broader interpretation, that is, as embodying nature and that which ignorance dominates super nature, then he should no longer be afraid of another or of himself. However, the wisdom of the infinite shall find expression and the great time limitor shall weigh in the balance, that is, Saturn, and reject time and time again, and then again soon after aeon, until ultimately even that grim arch restrictor shall permit his pupils to enter the new path in silence from which man's own ignorance was the only bar.

CHAPTER XXIII

---

*Aspects, Radical and Derivatives—An Aspect Necessi-
tates the Manifesting of Two Influences; these
Two Blended Give the Nature of the Aspect.*

In considering aspects and their nature, it will be found
all radical aspects are ordinate, that is, with the exception
of derivative aspects, which are formed by parts of two
radical aspects. Thus all radical aspects are intrinsically
able to form a regular sided figure which may be in-
scribed in a circle, the angles of the said figure touching
each one the circumference of such circle.

The word or term aspect is derived from the Latin
aspectus from *aspicio,* meaning, I look on, and the reference is to the astrological manner of supposing, truly
enough indeed, that planets posited at certain distances
of longitude look upon one another or receive and distribute one another's special influence more readily and
with greater power than when situated at distances
apart, the joining line of which to their centers, considering them as geometrical points having no objective existence, does not form the side of an equilateral circumscrib-
able figure.

The geometrical aspects are best illustrated by the performance of the problems in geometry, which have reference to the construction of regular sided figures within a
circle. The circle is ever considered as being composed
of three hundred and sixty degrees, this being the approximate number the Sun travels in a year through the
Zodiac.

It must be kept in mind from the beginning, that each
separate aspect is an angle of a certain number of degrees,

depending upon the varying planetary situations in the Zodiac, and the particular reasons of the general credit, or discredit, or strength, or power of a planet's ray has been, and is today with the student of this divine science, an unsettled question, that is where this ray or aspect or influence has its origin. It is conceded that a certain force does exist and is able to distribute its power through millions of miles of space, but the manner of how it is generated and by what method is an open question.

Ptolemy held that aspects made from the signs, that is, masculine and feminine, agree and transmit favorable influences, while those aspects formed from signs of an opposite nature disagree and are the occasion of discord. However, this idea does not hold good in the fact of the opposition, which aspect is ever formed from like signs, that is, either both masculine or both feminine, and the opposition is considered a very evil aspect. We have learned that in judging of the nature of the aspect, the esoteric nature of the signs must be considered, then apply the law of attraction and repulsion and we find the primary cause of aspect. An aspect necessitates the manifesting of two influences. These two, blended, give the nature of the aspect.

We may consider the Zodiac as represented by a vast ethereal ocean, as did the Egyptians, who have symbolically typefied the Sun's passage through the signs of the Zodiac as a ship passing over the bended form of a woman whose robe is studded with stars, while beneath reclines another form of horns, which represents the Earth in space and surrounded by a vast ethereal ocean or belt of influence. In the crypt of an ancient church at Piacenza is to be found a mosiac pavement before the altar upon which water is symbolically represented by waving lines representative of the sign Aquarius, and with fishes swimming therein representative of the sign Pisces, also plaques containing the Zodiacal signs, the two signs Aquarius and Pisces symbolizing the two last and most perfected signs, the whole representative of the Zodiac as a great ethereal ocean encompassing the Earth.

In this Zodiac move the Sun and the planets, the Sun, we learn, never removing its track or orbit from the central line or the ecliptic, while the planets occasionally temporarily leave it, therefore producing latitude either north or south, as the case may be, while the Sun, remaining as it does upon the line of ecliptic, is said to possess no latitude.

We may say here that, in considering aspects primarily, the signs of the Zodiac and the constellations by the same name must not be confused as being one and the same. Ages ago the constellations occupied the signs now bearing their names, but by the precession of the equinoxes this is now changed. For instance, the sign Aries changes its place backwards fifty seconds in space yearly, and the plane of the equator crosses the ecliptic twenty minutes sooner in time so that the equinoctial point retrogrades one degree every seventy-two years. As a consequence this makes a variance in longitude and latitude of the stars, necessitating a revision in celestial globes, atlases, etc., every seventy-two years.

The first point of Aries is now in Pisces. The reason this change of constellations does not affect the thirty degrees section into which the Zodiac is divided, is that an occult property belongs to each section, a property quite inherent, unchangeable, and in no wise dependent upon the groups of fixed stars which bear the same names as the signs in question.

---

*The Zodiac is a Vast Sea of Wonderfully Mobile Ether,
    Utilizing the Planets as Mediums Between It and
    Things Mundane.*

In considering the constellations and Zodiac as separate influences, if it can once be understood that actually the Zodiac is a vast sea of wonderfully mobile ether impressed universally with specific virtues and utilizing the planets more or less as mediums between it and objects or things mundane, the difficulty will have vanished and the problem is clear.

Aspects Themselves Are Never in a Straight Line.

As regards aspects, it is necessary to obtain a clear understanding of this subject before entering the more complex subject of directions. It must be kept in mind that aspects themselves are never in a straight line, as can best be illustrated by drawing a circle, inside of this circle draw a hexagon or six-sided figure;* and for instance we wish to illustrate the aspect known as the sextile and formed when planets are two signs or sixty degrees separated, or what we may better term here an angular arc of sixty degrees, and which comprises one-sixth of the complete circle of three hundred and sixty degrees; a line may now be drawn representing the ascendant, while a triangle may be drawn first a line from the twelfth house where the line intersects with the circle, to the second house, where the cusp of the second house intersects with the circle. This will form the base of the triangle. Then in the center of the figure a small circle with the cross may be drawn. This represents the Earth. Then by drawing two perpendicular lines from the points of the base to the Earth, which forms the apex of the triangle, it is clearly perceptible that the true sextile aspect is the measure of the intercepted arc between two angle points, which are at the two points of the base of the angle at the cusp of the twelfth and second house, while the inci-

---

*See next page.

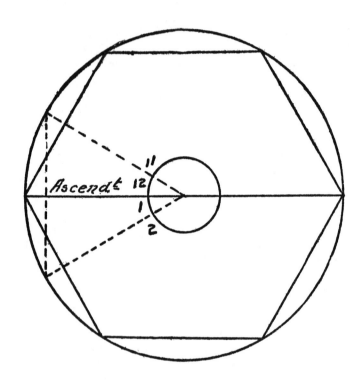

dence of the rays are at the apex of the triangle, or the Earth, where the power would manifest and where its influence would be felt.

Thus stars and planets, being posited that the arc lying between will equally divide the great circle into six parts, are then considered to be in sextile aspect to one another, and the same rule applies in proportion in other aspects.

It will now be seen by looking closely that three distinct series of progressions or affinitive aspects may be formed; first, an even series divisible by six, second an even series divisible by five, and an odd series divisible by five. While the first two form an even series, the third series are odd. Taking the first series, we have under this head aspects separated by eighteen, twenty-four, thirty-six, seventy-two, one hundred and eight, one hundred and forty-four and one hundred and fifty degrees. Under the head of second, we have aspects formed by thirty, sixty, one hundred and twenty, and one hundred and fifty degrees; under the third head, we have aspects formed by forty-five, ninety, one hundred and thirty-five and one hundred and fifty degrees. The first two series are considered favorable, while the latter is unfavorable. The aspects comprised in the first series are little used for the reason that they are comparatively modern additions to the list of aspects, and their influence is not of so great importance as the second or third series. They act on the physical plane and have little to do with the ordinary every day life. For this reason they are only transitory and tentively appreciated and appertain to the occult. The second and third series ever act more or less powerfully on the material plane, casting their influence for good fortune or the contrary, as the case may be. The second series comprise those aspects indicative and productive of pleasure and joys of life, while the third series comprise those aspects productive of indignities, the pains and sorrows of the earth life.

We have learned that the first series are more associ-

ated with the psychical than the physical. The two last
series, composed of sextile, trine, etc., and of the square,
opposition, etc., are associated with material influences.
This first and second series were considered by the Oriental to deal out the worldly rewards and punishments
that the ego has earned in its manifestation up to the
higher expression of individual consciousness, and were
not to be regarded merely as the results of accidental
positions of the heavenly bodies at the birth or the re-
birth of the ego; thus considering the horoscope to be a
species of passport accurate, or a signboard in which the
past expression may be read by a due appreciation of the
merits and demerits intended for the present.

In considering the parallel, it must be judged in the
same manner as the conjunction, that is, of good import
when planets of sympathy to one another are concerned,
while unfavorable if there exists no sympathy between
them, or when so-called evil planets are concerned. It
will be observed that the second series of aspects is built
on the triangle form, while the third, or odd series, is
built or formed on the cube, and·inasmuch as the Pythagoreans maintained in the first of the five regular solids,
that is, the tetrahedron or pyramid, the tetractys is to
be found that a point answers to unity; a line to number two; a superficies to number three and solidity to
number four. The student may form some conception of
the bonds which the two last of these series of aspects
hold to one another while remaining free enough to
operate in seemingly opposite directions, or rather by
opposite methods, as for instance, the number three expresses the triangle in the ternary nature of its sides and
angles and is looked upon as dissociated units; they are
triple, while together they constitute a unit, that is, the
synthesis of the whole angles and sides forming the containing figure.

This is what may be termed completing the triune by

the quarternary.   The angles of a triangle, it must be kept in mind, are ever equal to two right angles.

In a further consideration of these numbers, the student will observe that the indicative power is fourfold and consists of thought or mind science, opinion and sense, or in physics, metaphysics, ethics and theology.   Truth was considered as a universal measure and the idea was held by some of the learned teachers that the tetrad, or number four, typefied the father himself, as the number four contains the decad, that is, the finite interval of number, as before the ten is completed the sum of the first four presents all the virtue of the ultimate decad in this wise, one plus two plus three plus four equal ten.   The number four is the arithmetical mid-point between one, the unit, the infinite, and the number seven, which we have learned represents the Virgin, inasmuch as it produces no number between itself and the number ten.   The number eight is the reflection of number four, and one plus two plus three plus four plus five plus six plus seven plus eight equals thirty-six, the number of decanates of the Zodiacal circle, also representing the letters of the tetragrammaton and the numbers of the sephiroth.

In preparing the way for the student to fully grasp the mode of directing stars and planets, it will be well to study well the higher esoteric principles of numbers, as has been given heretofore in former lessons.   It is absolutely essential in dealing with this subject and a knowledge of the esoteric is necessary to a full comprehension of this science.   In modern research there has been too much dealing with exoteric only.

CHAPTER XXIV

Demonstrating the Use of the Polariscope.

In demonstrating by experiment a polariscope will be of great assistance in obtaining a clear understanding of this subject, as for instance, take the polariscope. Let the source of light be the Sun or diffused daylight, or the light from some other source may be utilized, provided with a ground glass or other equivalent shade, the polarizing plate being properly adjusted and the index placed at zero of the upper graduated circle. A film of selenite may be placed in front at the proper focus when its image will be observed to be of a tint varying with the thickness of the film. Now, if this be slowly revolved the color will gradually disappear until it will vanish entirely. However, if the revolving process is continued the color will gradually appear once more and will regain its maximum brilliancy. Thus two points will be gained or observed in the evolution, one where no color appears whatever, and the other where it is the most brilliant, and these two points are important here as the point of greatest chromatic intensity and the vanishing points are found to be inclined to each other at an angle of forty-five degrees.

Then adjust the instrument again, arrange the film of selenite and allow it to be of such thickness that its tint appears red in color and when its red image is visible in the analysing plate slowly revolve it, at the same time observe carefully the arcs of rotation upon the graduated circle. It will be observed that the red color gradually diminishes in brilliancy until, by the time the analyzing plate has moved through an arc of forty-five degrees, it will have disappeared entirely. Then by further revolution the film will slowly assume a green tint that is the complement of red and will attain its maximum brilliancy at ninety degrees. Then during the next further revolution of the next arc of forty-five degrees, the green will have vanished entirely and when one hundred and thirty-five degrees have been passed the red will gradually appear again and finally attain its most brilliant hue at one hundred and eighty degrees. Then by further process of revolution the same phases are repeated through arcs of forty-five degrees until the complete circle has been made through the three hundred and sixty degrees, and if the film of selenite is of such thickness as to give other tints besides those mentioned, their complementary colors will also be found, that is, the colors obtained at zero, at ninety degrees or at one hundred and eighty and two hundred and seventy degrees, will be invariably such that being united they would constitute white light. We will consider this subject further at a more opportune time when considering colors and their relationship to points of astrological science. However, this experiment will illustrate to a great degree the complementary natures of aspects and their mechanism, and while it is observed that the position of greatest intensity in an aspect is at its exact formation, such formations being seldom found in the horoscopal figure. It is important to consider the ratios of the gradually lessening or increasing activity of the force from or up to a certain known point, for it must be kept in mind that in the Zodiacal degrees there are no blanks to be found, and the same is true with regard to the circle of aspects, but only an appearance there, to be explained on a rational basis. It will be observed that everything is continuous. There is no hitch or flaw, but ever an undeviating progression along lines that are fixed and immovable.

Separate and apart from the undulating lines of aspects that are fixed and immovable, there inheres a particular nature to each several aspect. The modern student is too apt to make the error of lumping terms, sextiles, etc., together and denominating the same with the squares and oppositions, and calling them evil. However, when this method is pursued, the fact is willfully ignored that each and every aspect possesses a special character and influence of its own which can be clearly demonstrated in each particular case. The general efficacy of each and every aspect arises from its harmonical proportion and bears equally on the laws of light, color and harmony as on those of astral science, and also every agent acting by itself does so according to its own peculiar form and virtue. It will therefore be clearly understood first of all, that aspects are not active forces, but are rather passive mediums or agents when being formed, will thus enable the celestial bases of influences proceeding from the heavenly bodies to function. They do, in fact, unlock the sluice gates through which the forces rush, in just the same way as at the moment of birth, with the exception that the ego can attach itself to the clay with which the stars and planets are in sympathy.

In considering directions we find there are two kinds which are best termed mundane and Zodiacal. Primary directions really belong and come under the head of mundane directions, and what is generally termed secondary directions come under the head of Zodiacal directions, that is, to be more clear, the latter is made according to the circle of the Zodiac, while the former is made according to that of the Earth or in respect to the mundane distances of houses and angles.

Primary arcs are constituted by the diurnal motion of the Earth immediately preceding and following the birth, either by aspect or body, and it must be conceived from the beginning of calculations that the planets' places at the moment of drawing into the lungs the first astral breath only are to be dealt with. They have all the efficacy impressed upon them of the heavenly body which has moved away, therefore it will be observed that the primary directions for a lifetime in the physical are formed within a very few hours after the birth takes place. It may be either by body or by aspect to angles or to other planets or to a planet's radix or original place at birth. These are then considered the most powerful agents in the production of events, though their calculation involves great patience on the part of the student and also a certain amount of mathematical ability, two essentials that are not always at the disposal of the student, but by care it is not difficult to develop these. Each student that becomes interested in this divine science will possess some essential qualities that will enable him to better interpret some branch or line of the science better than another, and one of the first essentials is for the student to become thoroughly familiar with his or her own natal chart and be able to read therein without prejudice or bias his or her own nature, character and natural tendencies. When the student has accomplished this he is then better able to interpret and render judgment in the horoscope of others. First know thyself is the great teaching.

The secondary system, primarily intended as an aid in the foregoing, is purely Zodiacal, and was in great favor with the Hermetists and Chaldeans, also the Arabians, hence it often goes by those names. It is based upon the daily motion of the Sun, Moon and planets subsequent and antecedent to the birth, and possesses undoubted power and influence. It is a method that should appeal to the modern student of this divine science.

Rule Given for Obtaining the Mid-distance of any Planet.

You will be able to obtain the mid-distance of any given planet by following the rule as given. Note whether above or below the earth, and if posited nearer the mid-heaven or nadir, ever keeping in mind that computation must be made in order of signs and in finding the mid-distance of Neptune you will observe that Neptune in the seventeenth degree of the sign Aries, near the cusp of sixth house, gives the right ascension sixteen degrees and five minutes and as the right ascension of nadir is three hundred and twenty-nine degrees and fifty minutes, naturally the degrees of the circle must be added in order to obtain the correct mid-distance.

The semi-arcs are to be obtained by adding the tangent of the latitude of birth to the tangent of declination of the star or planet in question. The result obtained is termed the cosine of semi-arc. There are various methods of arriving at this result. Now, if the latitude of the birthplace and the declination of the planet are the same, that is, both north or both south, the result is the semi-nocturnal arc, and if the diurnal arc is required it must then be subtracted from one-half of the circle which we know to be one hundred and eighty degrees. Then again, if the declination of the star or planet and the latitude of the birthplace from which the figure is erected are not of the same, that is, both north or south, one being north the other south, or vice versa, the result obtained by adding the tangent of latitude of birth to tangent of declination of planet will be the semi-diurnal arc. It will be observed that the opposite arc is ever the complement necessary to form the half-circle comprising one hundred and eighty degrees. Therefore, the student will observe that when an arc of either name be given, that is nocturnal or dirunal arc, in order to find the other the result obtained must be subtracted from the half-circle, that is, one hundred and eighty degrees.

We may now consider an example to better illustrate this point. We desire to know the semi-diurnal arc of the Sun in the nativity as given.

First, the tangent of the latitude of place of birth is

fifty-two degrees and twenty-eight minutes north, which
equals ten and one hundred and fourteen and four hundred and ninety-seven thousandths.   We observe the
tangent of the declination of the Sun is twelve degrees and
forty-four minutes north.   This equals nine and three
hundred and fifty-four and fifty-three thousandths.   The
result after adding gives nine and four hundred and sixty-
eight five hundred fifty thousandths, equals seventy-two
degrees and fifty-four minutes, which is the cosine of the
semi-nocturnal arc of the Sun.   In this example we find
the declination of the planet in question is the same as
the latitude of birth figure, that is, north.   Naturally in
our hemisphere the latitude will ever be north.   The
result obtained is the semi-nocturnal arc, and as we desire
to know the semi-diurnal arc, we must go further and
make the subtraction as given heretofore, as, for instance,
one-half the circle equals one hundred and eighty degrees.
We have obtained the semi-nocturnal arc, seventy-two
degrees and fifty-four minutes, which amount subtracted
from the half circle gives one hundred and seven degrees
and six minutes, the remainder which represents the Sun's
semi-durnal arc.   The speculum being completed, the nativity stands in a position for directing and do this accurately some systematic attention must be given to the
order in which the aspects are, progressively, from the
earliest period of infancy to the latter period of old age.

The speculum is a table made out clearly showing the
elements as required of each planet in the figure.   In
directing the stars and planets, system must be used and
attention given as heretofore stated to the order in which
aspects are formed progressively from the moment of
birth to the latter period of old age.   However, this may
not be rigidly followed out for a collection of the arcs
in proper order may be made after all computations are
completed.   The better method is to begin with the various directions to the angles in mundo in as near the
natural formative procession as is possible and then fol-

low these with the mundane directions among the planets themselves, keeping in mind the importance of the Rapt parallels. Next may be considered the Zodiacal aspects to the angles, and then when this is completed the mutual configurations of the planets may be obtained.

A little careful study and practice will enable the student to grasp the necessary details and work out the figure in these various orders, that is, the most essential part in the primary system easily and with pleasure, for it becomes intensely interesting as the student is able to grasp fully the workings of the methods of directing.

The directing to the part of fortune, fixed stars Caput and Canda Draconis or the Dragon's Head and Tail, as well as to the cusp of houses are not so important in the natal figure and are required more directly when dealing with the mundane or comparing a natal figure to the mundane configurations, and if the student has a desire to investigate these and other points in the science of Astrology, he should at least be able to perform the problems necessary in the computations of the directions, beginning with the various directions to the angles in mundo as in the natural formative procession as stated. It is well to bring out some examples of mundane and Zodiacal directing, for this will give a more clear understanding to the student, who may form some idea of the practical processes concerned, as well as the theoretical reasons for them, and thus be enabled to follow the explanations as will be given. In the first place this science is so vast that there is scarcely time to extenuate greatly the various problems, and then, too, there are so many and various methods of working that an example of all would take up too much valuable time. We will, therefore, follow out one method and system which will best answer our purpose and is best adapted to the student of this age.

To begin with, the student will observe that aspects in mundo are measured by the semi-arc of the promittor and for those in which the mid-heaven is concerned we may employ right ascension, while those to the ascendant are worked out by semi-arc of the planet or body moved. Thus by attending to the following method all directions to the angles in mundo completed in an ordinary lifetime may be completed easily in something less than thirty minutes' time, especially by one who is naturally adapted to the work. It will be observed that after the mid-distance is obtained in the speculum you also have the conjunctions and oppositions to the mid-heaven as well as the mundane squares to the ascendant, for the reason that when a planet is either on the cusp of the tenth, the mid-heaven or the cusp of the fourth or nadir, it is then in exact square to the planet or body on the cusp of the first in mundo; that is, if a planet is there and it is readily observed that the cusps of tenth and fourth are in exact opposition to one another.

After having obtained the mid-distance of the stars and planets in the speculum and the conjunctions and oppositions to the mid-heaven, also the mundane squares to ascendant, then the several aspects to the mid-heaven and the ascendant and first house may be obtained therefrom by aliquot portions of the various semi-arcs. In obtaining these the student will begin with those aspects to the mid-heaven, premising that when on the completion of the aspect the planet or star is below the Earth the semi-nocturnal arc is utilized, though when above the Earth then the semi-diurnal arc is used. The student must give this point particular study, as for instance, in the nativity as exemplified when the Moon arrives on the cusp of the eleventh house it will then form a semi-sextile to the mid-heaven and at the same time will form a sextile aspect to the ascendant. Thus by one computation the two aspects or directions are obtained, both being calculated by the diurnal arc. When the planet Saturn passes on to the cusp of the second house the student will observe that Saturn is then in trine aspect to the mid-heaven. However, this planet being below the Earth the student must necessarily utilize the nocturnal arc in obtaining direction of same. In the speculum the student will observe the semi-arcs are for the radical positions of the planets, that is, in all instances where the planet in question is below the Earth the arc as set forth in the column will be its nocturnal arc, and when above will be its diurnal arc, and from the rule as given either arc is easily obtained.

It will be well here to give a table of complementary arcs to the mid-heaven and the ascendant in mundo, which will show to the student at a glance what synonymous aspect is formed to the ascendant, if any, and when a star or planet is directed to an aspect of the mid-heaven in mundo or vice versa. First we will take the aspects as formed below the Earth to the mid-heaven, the trine to mid-heaven, that is four houses or signs, or one and one-third semi-arc; the complementary arc forms a semi-sextile to the ascendant. Then the sesquadiate aspect to mid-heaven, being four and one-half houses or signs or one and one-half semi-arc, the complementary arc forms a semi-square to the ascendant; then the aspect formed to the mid-heaven of one hundred and fifty degrees removed or five houses or signs apart and is one and two-thirds of the semi-arc; the complementary of this aspect forms the sextile aspect to the ascendant. Then we have the opposition aspect to the mid-heaven, or six houses or signs separated, or a whole arc. The complementary arc of this aspect forms the square aspect to the ascendant, as we have stated heretofore.

We may next consider the aspects as formed above the Earth. First, the conjunction aspect, being in the same degree and sign and the complementary arc forms the square aspect to the ascendant. As will be observed, the mid-heaven or cusp of tenth is ever in square aspect to the ascendant and a planet here in conjuction would naturally be in square. The semi-sextile aspect to the mid-heaven one sign or house apart or one-third semi-arc; the complementary of this arc forms the sextile aspect to the ascendant; the sem-square aspect, one house and one-half apart or one-half semi-arc; complementary arc forms semi-square to ascendant.

When the planet or star is found in sextile to the mid-heaven, that is, two houses or signs separated, or as two-thirds of the semi-arc the complementary arc will then form a semi-sextile to the ascendant. When the planet is found in square to the mid-heaven by direction, that is, three houses or signs separated, or one whole semi-arc, the complementary arc will then form a conjunction to the ascendant. The remaining aspect will be found to be independent or of minor importance.

In order to more clearly illustrate the directing of planets it will be well here to direct the mid-heaven and ascendant to the aspects as just given of the ,planet Mars in the figure given.

First, we find Mar's mid-distance or Mars to mid-heaven seven degrees and five minutes; mean conjunction and also squares the ascendant one-third of the semi-arc which is diurnal, being above the Earth, equals thirty-seven degrees and fifty-four minutes. We subtract the mid-distance of Mars from the semi-diurnal arc. This leaves a remainder of thirty degrees and forty-nine minutes. This is the mid-heaven directed to the planet Mars. We find the distance is one sign separated, or one-third arc, which gives a semi-sextile aspect of Mars to the mid-heaven and at the same time forms the trine to the ascendant. We then add to thirty degrees and forty-nine minutes one-sixth of the semi-diurnal arc of Mars, which is eighteen degrees and fifty-seven minutes, gives the mid-heaven semi-square Mars forty-nine degrees and forty-six minutes. ,This also gives a sesquadrate aspect to the ascendant. To this we then add one-sixth of the semi-diurnal arc which place gives a semi-sextile aspect to the mid-heaven, that is, sixty-eight degrees and forty-three minutes. This place will also be found to be one hundred and fifty degrees removed from the ascendant. Then, again, to sixty-eight degrees and forty-three minutes we add one-third of the semi-diurnal arc, that is, thirty-seven degrees and fifty-four minutes gives one hundred and six degrees and thirty-seven minutes. Now when Mars arrives here at this degree in the figure by referring to the table it will be seen that Mars is in square to the mid-heaven, the complementary arc of which forms an opposition aspect to the ascendant. Then to one hundred and six degrees and thirty-seven minutes we add one-third of the semi-nocturnal arc that is, twenty-two degrees and six minutes. We must add the semi-nocturnal arc for the reason that we are now below the cusp of the descendant or cusp of the seventh house and also below the Earth.

We find that when Mars arrives at one hundred twenty-eight degrees and forty-three minutes it will then be in trine aspect to the mid-heaven or mid-heaven will then be in trine aspect to Mars, and one hundred and fifty degrees from the ascendant. To this last number of degrees and minutes we add another one-sixth of the semi-arc nocturnal, that is, eleven degrees and three minutes. This totals one hundred thirty-nine degrees and forty-six minutes. This gives the mid-heaven in sesquadrate aspect to Mars also in sesquadrate aspect to ascendant. Again we add one-sixth of semi-nocturnal arc. This gives one hundred fifty degrees forty-nine minutes. Mid-heaven is then one hundred fifty degrees from Mars and trine to ascendant.

To one hundred and fifty degrees and forty-nine minutes we now add one-third of the semi-nocturnal arc, that is, twenty-two degrees and six minutes. This gives one hundred and seventy-two degrees and fifty-five minutes, from which point the mid-heaven will be in opposition aspect to the planet Mars. The complementary forms the square aspect to the ascendant.

In a careful study of these directions it will be observed that the proportional parts of the semi-arcs have been employed according to the table as previously given, and in a similar manner any other desired aspect may be computed by taking the aliquot part in accordance therewith. It will also be observed that the whole semi-arc lies between the cusp of the tenth house and the cusp of the seventh house or descendant, that is so far as the Mars aspects in the above calculations are concerned, and one-third of this semi-arc will then be equivalent to one mundane house or thirty degrees and two-thirds of this semi-arc will be equivalent to two houses or sixty degrees, while one-sixth of the semi-arc will be equivalent to one-half of a house or equal to fifteen degrees. Therefore, it will be clear to the student why this method of calculation is used.

Then, again, it will be observed that from the cusp of the ninth house to the mid-heaven or cusp of tenth house is one-third of the semi-arc which measures thirty-seven degrees and fifty-four minutes; therefore if the planet Mars were exactly at the zenith this would then be arc of direction of the mid-heaven semi-sextile to Mars or the ascendant in trine aspect to Mars. However, the student must keep in mind that allowance must be made for the number of degrees that Mars is deflected from the tenth cusp and this amount must be taken from one-third of semi-arc or thirty-seven degrees and fifty-four minutes in order to find the true measure, as is demonstrated in the previous working out of Mar's aspects.

The student should carefully observe what occurs after he passes from the semi-diurnal arc to the nocturnal arc. The square of Mars and the mid-heaven, that is, it amounts to the same, and the opposition of Mars to the ascendant has been formed and consequently to proceed in the order of the lengthening of the aspects it becomes necessary that the planet move below the Earth in order to form the succeeding aspects, that is, when the descendant is reached in the calculations we know this point is

ever in square to mid-heaven and in opposition to ascendant, and from here on the nocturnal arc is used and continuing as before.   Naturally before the square has
formed to mid-heaven in this case the native would, no
doubt, have left the mortal form, but this fact would not
invalidate the object for which we have worked out this
calculation, that is, to demonstrate how one aspect may
be obtained from a foregoing one in unbroken continuity.
The whole arc stretches from the cusp of the tenth house
to the fourth or nadir and should consist of one hundred
and eighty degrees, neither more nor less.   Thus the
opposition of Mars to M. C. should equal this number
minus its distance from cusp of tenth, that is, Mar's mid-
distance plus arc of direction, opposition, mid-heaven
should both sum to one hundred and eighty degrees.   We
may test this arc of direction of Mid C. opposition to
Mars equal one hundred seventy-two degrees, fifty-five
minutes, plus mid d. Mars seven degrees and five minutes
gives half circle or whole arc one hundred and eighty
degrees.   It is, therefore, possible to test all aspects in
this same manner.

The student will observe that all the aspects of the
planet Mars in the figure were formed after birth by
mundane motion, that is, that motion which the planet
would appear to have if considered for a few hours, as
for instance, would the Sun or Moon, their rising, cul-
mination and setting which the student knows occurs
daily and irrespective of their positions in the Zodiacal
signs; therefore, it will be observed that the conjunction
of the planet Mars and the mid-heaven took place shortly
before the birth, and if the mid-distance of Mars, that is,
seven degrees and five minutes, be taken as the directional
arc, it must be so understood, for Mars will not travel
back to the mid-heaven, the daily motion of the Earth
never varying its direction of rotation for this reason,
because such aspect as the conjunction of Mars and the
mid-heaven could not be formed after the birth.

Many students might not consider the effect and reprehend the calculations of such positions, but experience will demonstrate to the student what he may expect, and from this must his decisions be made. It will be observed that the aspects to the ascendant may be similarly calculated. The student must, however, keep in mind that in doing this that many of these aspects will, no doubt, have been included by complement; that is, when computing these aspects to the mid-heaven. It will be observed that the mundane square to the ascendant is ever the planet's meridian distance, either above or below the Earth, while the opposition, as likewise the conjunction, is ever in square to the mid-heaven. These positions will in a little time become familiar to the student. It will be well to give much practice in bringing arcs to the angles in mundo before trying to go further in the calculations in order to obtain a thorough comprehension of the process of computation, as well as the reasons for same, and when the student has progressed sufficiently to thoroughly understand this he may next proceed to the computation of the mundane directions of luminaries.

As the student will observe, these are based upon certain relative distances from angles and the cusps of houses and depend for their constitution and dissolution on the diurnal motion of the Earth. A direct direction is one in which one of the luminaries either the Sun or Moon is supposed for the time to be stationary, that is mundanely and to which other planetary bodies exert completed rays which are formed successively by their apparent motion from east to west; therefore, it will be observed the Zodiac is not requisioned, as for instance, in the figure as given for example, the Sun being supposed to be stationary in its relative position to the western cusp or rather the Sun's mundane place being understood to retain the radical solar power, for it must be kept in mind that the body of the Sun itself moves onward in continuity. Saturn rising in diurnal conformity will form an opposition aspect in the twelfth house at an equivalent distance from the eastern angle as the Sun is from the western angle. Therefore, the problem resolves itself first into a question of learning the radical distance of the Sun from the cusp of the seventh house. The next step will be to determine the distance Saturn must travel in order to arrive at the same distance of the other side of ascendant. This will be the arc of direction.

When Saturn has arrived at the same proportional distance above the ascendant as the Sun is below the western angle or descendant, this will constitute the arc of direction, that is, not to the Sun's place as it will actually be at that time, but to its former position in the radix in which it was posited at a critical moment, and that moment was the moment of the birth of the soul into physical expression, and at this highly psychic time the specific irradiations of a star or plane becomes similar to a photographic negative or plate. What is then manifested or reflected remains fixed and efficacious. We may now consider this arc of direction as an example, that is, we wish to learn the arc of the Sun in opposition to Saturn in mundo by direct direction as the semi-arc of the Sun proportional logarithm equal nine and sixty-seven hundred forty-six thousandths, this is to the Sun's distance from cusp of seventh, one and ninety-nine hundred sixty-four thousandths. The semi-diurnal arc of Saturn equal forty-seven seven hundred seventy-two thousandths, to the second distance Saturn nine degrees and twenty-one minutes, equal one and twenty-eight four hundred eighty-two thousandths. Then the distance of Saturn from the cusp of the first house added to the second distance or proportional distance Sun is from cusp of descendant nine degrees twenty-one minutes, gives the arc of direction Sun in opposition Saturn direct direction in mundo as fifty-five degrees and ten minutes as the distance Saturn must travel to in order to form the opposition aspect. The distance of the Sun from the cusp of the descendant is found by taking the difference between its semi-arc and mid-distance, while that of Saturn is found from the ascendant by similar means. However, as the planet Saturn forms the opposition above the Earth, its semi-diurnal arc must be taken, and this rule applies in all cases where the directed planet crosses the horizon from below, though had the position of these two planets been reversed in the figure, the Sun being above the cusp of ascendant and Saturn on the cusp of the eighth house, in order to complete the opposition aspect, Saturn would descend into the sixth house and then in the place of working with its radical semi-diurnal arc it would have been necessary to use the semi-nocturnal arc. This point must be kept well in the mind of the student. It is quite important. It will be observed that the primary and the secondary distance of Saturn have been added together in order to obtain the directional arc according to this rule, the bearings of which will be seen clearly as we proceed. Thus if the secondary distance of the planet be upon the side of the cusp whence primary was taken when aspect was complete, subtract primary and secondary distances; but if otherwise, add them.

We may now consider another example directing the Moon to a square of the Sun by direct motion in mundo. We have found the semi-arc of Sun in proportional logarithm to be nine and sixty and seven hundred forty-six thousandths, the distance of Sun from descendant in proportional logarithms one and nineteen nine hundred sixty-four thousandths, semi-arc of Moon ninety degrees twenty-five minutes equal twenty-eight and twenty-two thousandths, to secondary distance of Moon fourteen degrees forty-three minutes equal one eighth and seven hundred thirty-two thousandths, then secondary distance of Moon added to Moon distance from cusp of tenth give the arc of direction, Moon square to the Sun.

In obtaining the arc of direction of Moon in square to

the Sun by direct motion in mundo, it must be observed
that in case the Moon moves forward through the tenth
house, crosses the cusp of tenth and partially traverses
through the ninth house ere it reaches the point where
it forms the square aspect, and therefore both primary
and secondary distances taken as they would be from
opposite directions or sides of the cusps, the two distances must be added together.

We may now give some attention to what is termed
converse directions. In considering these directions the
luminary itself is moved forward until the required aspect
is formed with the place of the promittor, and, as will be
observed, this will necessitate a variance in the method
of working. For instance, as the semi-arc of a fixed
planet is to its distance from the nearest cusp, so is the
semi-arc of the body directed to the second distance of
same. We may consider an example here to better illustrate this point and make it perfectly clear. We will
direct the greater luminary by converse motion in mundo
to a conjunction of the planet Neptune, using the figure
as first given; thus, as the semi-nocturnal arc (it must
be kept in mind that we are now working below the cusp
of descendant, therefore we use the nocturnal arc) of
Neptune, that is, eighty-three degrees and sixteen minutes, the proper logarithm, equals nine and sixty-six and
five hundred twenty ten-thousandths; as this is to the
distance of the planet Neptune from cusp of sixth, that is,
nine degrees fifteen minues, which equal in logarithms
one and twenty-eight and nine hundred thirteen ten-
thousandths, so is the semi-arc of the Sun, that is, seventy-
two degrees and fifty-four minutes, which equal in proper
logarithm thirty-nine two hundred fifty-four ten-thousandths; this is to the secondary distance of the Sun, that
is, eight degrees and six minutes, equal one and thirty-
four and six hundred eighty-seven ten-thousandths, and
to form the conjunction the Sun must necessarily pass
the cusp of the sixth house, from which its primary dis-

tance may be easily reckoned as twelve degrees and fifty-six minutes. Thus, as has been stated in the problem just considered, the sum of primary and secondary distances is to be taken for the directional arc, as, for instance, the Sun's primary distance from the cusp of the sixth house is twelve degrees and fifty-six minutes. The secondary distance of the Sun we have found to be eight degrees and six minutes, which, added together, gives a total of twenty-one degrees and two minutes. This sum constitutes the arc of direction of Sun in conjunction to planet Neptune, converse in mundo. The student will observe that in order to find the distance of the Sun or any planet or star from the sixth cusp when planet or star is below the Earth, it is obtained by taking two-thirds of its semi-nocturnal arc from its mid-distance. In further consideration of this direction, we find that the Sun arrives at the radical place of planet Neptune by the diurnal motion of the Earth just in the same manner as in the direct direction as Saturn was shown to do when directing said planet to the opposition of the Sun in mundo; and further, as the Sun crosses the cusp of house from which primary direction was taken, being the sixth in this instance, the primary and secondary directions must necessarily be added together.

Let us now consider another example in converse motion, as to direct the Sun to a square conversely in mundo to the planet Saturn, and we proceed in this wise: as the semi-arc of Saturn, that is one hundred and twenty degrees and five minutes, which in proportional logarithms equal nine and eighty-two four hundred twenty-one ten-thousandths, as this is to Saturn's distance from cusp of second house, that is, five degrees and forty-seven minutes, which equals in proportional logarithms one and forty-nine and three hundred and nine ten-thousandths, so is the semi-arc of the Sun, which equal in proportional logarithms thirty-nine and two hundred and thirty-four

ten-thousandths, and this is to the secondary distance of
the Sun, that is, three degrees thirty minutes, which equal
one and seventy and nine hundred eighty-four ten-thousandths. Thus the Sun's primary distance from the cusp
of fifth house, which is thirty-seven degrees and fourteen
minutes, minus the secondary distance, which we found
to be three degrees and thirty minutes, and we have
thirty-three degrees and forty-four minutes as the remainder, and this number in degrees and minutes constitutes the arc of direction of the Sun in square to Saturn,
conversely in mundo. The student will observe in this
example that the Sun has to be brought to a distance
within the cusp of the fifth house proportionate to Saturn's distance inside the cusp of the second house, and
to find Saturn's distance from the cusp of second, two-
thirds of the semi-arc must be taken from the mid-
distance of Saturn, while the distance of the Sun from
the fifth cusp is found by taking one-third of the Sun's
semi-arc from its mid-distance. Then as the aspect is
completed on the same side of the cusp from which the
primary distance is taken, the secondary distance is subtracted therefrom in order to obtain the directional arc.
We may now consider an example in order to better
illustrate the method of obtaining one arc from another,
using the same computations for both or several. Thus,
in place of working out five different problems in order
to obtain the arcs of direction of Sun in semi-sextile,
semi-square, sextile, square and trine aspects, the student
may by the use of aliquot parts of the several semi-arcs
bring them all up in a much shorter time as well as in
more compact form.

As an example that will make this point clear, we will
direct the Sun to aspects just mentioned of the Moon by
direct direction. As the Sun's semi-arc, that is, seventy-
two degrees and fifty-four minutes, equal in proportional
logarithms nine and sixty and seven hundred forty-six
ten-thousandths, as this is to the distance of Sun from

cusp of the seventh house equal in proportional logarithms
one and nineteen and nine hundred sixty-four ten-
thousandths, so is this distance to the semi-arc of the
Moon, that is, ninety-four degrees twenty-five minutes
equal twenty-eight and twenty-two ten-thousandths.
This is to the secondary distance of the Moon, that
is, fourteen degrees and forty-three minutes, which
equal one and eight and seven hundred thirty-two ten-
thousandths. Thus, to form the trine aspect, the Moon
crossed cusp of eleventh house, from which its primary
distance is no degrees and five minutes; therefore to find
arc of one hundred twenty degrees the primary and secondary distance must be added together. When the trine
is obtained, one whole sign or house must be traversed
by Moon are the square aspect will be found, or one-third
of semi-arc.

It will be observed that while the aspects will decrease
in length owing to the positions of the two planets concerned, the arc of measurement will increase. Thus the
semi-arc to use in these proportional parts is that of the
body which is moved, which in this instance, as will be
observed, is the Moon. We may now proceed to obtain
the aspects in order. The secondary distance of the
Moon, we have learned, is fourteen degrees and forty-
three minutes; the primary distance from cusp of eleventh
house added to this, which is five minutes, gives a total
of fourteen degrees forty-eight minutes. This constitutes the aspect of Sun trine to the Moon. This is the
directional distance in mundo the Moon must travel in
order to form the trine aspect to the Sun. From this we
may easily learn the directional distance the Moon must
travel in order to find the remaining aspects of square,
sextile, semi-square and semi-sextile. For example, we
add to the directional distance of Moon trine to Sun,
which is fourteen degrees and forty-eight minutes, one-
third of the Moon's semi-arc, which is thirty-one degrees
and twenty-eight minutes. This gives a total forty-six de-

grees and sixteen minutes. This is the directional distance in mundo the Moon travels to form the square
aspect to the Sun. Then, to forty-six degrees and sixteen minutes we again add one-third of the Moon's semi-
arc, thirty-one degrees and twenty-eight minutes. This
gives a total of seventy-seven degrees and forty-four
minutes. This is the directional distance in mundo of
the Moon in sextile to the Sun. Then, again, in order
to find the semi-square aspect or forty-five degrees separated, we add to this number one-half of one-third of
the semi-arc, or one-sixth of the semi-arc, or fifteen degrees forty-four minutes. This gives a total of ninety-
three degrees and twenty-eight minutes. This is the
directional distance in mundo of Sun in semi-square
to the Moon. Then to find semi-sextile aspect we add to
this number one-sixth of the semi-arc, or fifteen degrees
forty-four minutes, giving a total of one hundred nine
degrees and twelve minutes. This gives directional distance in mundo of Sun semi-sextile to Moon, or the
distance Moon will travel in forming this aspect. The
reason for the use of these aliquot parts of the various semi-arcs has been explained heretofore. The principal point, for the student to keep in mind, is to use the
proper arc in calculating, that is, nocturnal or diurnal,
and converse directions may be obtained in exactly the
same way. It is very improtant for the student to keep
in mind that the semi-arc used is ever that of the body
directed, otherwise he will have much difficulty in his
calculations. Then taking the directional part of the
nativity in order, we may next consider the mundane
parallels, both direct and converse, and then later the rapt
parallels. Munane parallels are those equal distances
formed in the world from the upper and lower meridians
in contradistinction to those formed in the Zodiac by
being equally posited from the equator, as, for instance,
a star or planet posited on the cusp of the twelfth house
and another on the cusp of the eighth house represent

equivalent distances from the mid-heaven or tenth cusp; that is, so far as the world is concerned, and the same is true from the nadir or fourth cusp, though being removed from the former point by the space of two houses or two-thirds of the semi-arc diurnal, and must therefore be understood to be in mundane parallel.

The same is true of other distances, no matter whether they are formed from the upper or lower heaven in the same hemisphere or not, so long as the relative distances are the same, keeping in mind that if a star or planet has to pass the horizon to complete the aspect, the diurnal or nocturnal arc must be used, as the case may require. In working out these they will be found simple in calculation and in a little time the student will be enabled to proceed without any difficulty. We will now consider to direct directly, as, for instance, as the semi-arc of the Sun or Moon is to its mid-distance, so likewise is the semi-arc of the planet moved to its second distance, from which to find the arc the primary from the mid-distance or the nadir must be subtracted as follows: direct the Moon to the parallel of Mars direct distance in mundo; as the semi-arc of the Moon, that is, ninety-four degrees twenty-five minutes, the proportional logaritahm being nine and seventy-one nine hundred seventy-eight ten-thousandths is to the mid-distance of the Moon, that is, thirty-one degrees and thirty-three minutes in proportional logarithms equals seventy-five and six hundred twenty-seven ten-thousandths, so likewise is the semi-arc of Mars, that, is, one hundred thirteen degrees and forty-one minutes equal in proportional logarithm nineteen and nine hundred fifty-eight ten-thousandths to the secondary distance of Mars, that is, thirty-seven degrees and fifty-nine minutes, which equal sixty-seven five hundred sixty-three ten-thousandths. Then from the secondary distance of Mars, that is, thirty-seven degrees and fifty-nine minutes, subtract the primary distance from the mid-heaven, that is, seven degrees and five minutes. This gives a remainder of thirty degrees and fifty-four minutes, which is the arc of direction of Moon parallel to Mars in direct direction in mundo. It will be observed that in this instance Mars must be advanced until it arrives beyond the cusp of the ninth house, where the aspect will then be completed, so that the problem is to learn first the distance Mars must have from the cusp of tenth house to be the balance in mundane power or distance of the Moon's radical mid-distance; and second, the student must learn the intercepted arc between this, which may be termed the secondary distance, and the radical or primary position of Mars, that is, its mid-distance, while the difference will give the arc of direction. Now, converse directions will be found to have a little different method, though this is only apparently so, for in these converse directions the luminary's semi-arc occupies the third term of proportion, that is, the third term is the semi-arc of the moved or directed planet, be they converse or direct directions. For example, direct the Sun to the parallel of Saturn, converse in mundo as the semi-arc of Saturn, that is, one hundred twenty degrees five minutes is to the mid-distance arc of Saturn, that is, seventy-four degrees sixteen minutes, so is the semi-arc of Sun, that is, seventy-two degrees fifty-four minutes, to the secondary distance of Sun, that is, forty-five degrees and five minutes. Thus the primary distance of the Sun from nadir being sixty-one degrees thirty-two minutes, and from this we take the secondary distance of Sun forty-five degrees and five minutes, gives a remainder of sixteen degrees and twenty-seven minutes, which is the arc of direction of the Sun in parallel to Saturn converse in mundo.

CHAPTER XXV

The Science of Astrology Makes One Tolerant of All, Shows the Chains Under Which the World is Bound, also Giving the Means Whereby These Bonds May be Broken.

An understanding of this divine science makes one tolerant of all and shows the chains under which the world is bound, also giving the means whereby these bonds may be broken by pointing to a law of harmony that exists in the higher and perfect being pointing out the eternal justice of all manifestations, revealing the great truth that all mankind are to make themselves perfect in harmony with divine spirit. In realizing that there are no two individuals just alike, if we but pause to consider what this means, and by going still further in our analysis we find no two horoscopes that are just the same, each ego coming into expression under that particular stellar ray, the environment is colored to just that degree which it requires. This knowledge naturally calls for charity in criticising those who manifest follies. Investigation discloses with Mercury afflicted in a manner that will in time bring about the proper balance of that particular individuality. On the other hand, when purified senses and emotions are expressed in the horoscope, we ever find a pure and lofty intellect and moral growth of a high degree, leaving the desires of the objective behind to a greater or lesser degree, the mind reaching out toward the spiritual. True, in considering the horoscope of every individual, there is ever some cross to bear, some color vibrations that tend to mar the purity of the whole, though the more perfectly and harmoniously the character is balanced and consequently the natal chart, the more easily can the will find expression to overcome this imperfection, for it must be kept in mind that by will we choose our thoughts. In the time to come I feel confident that it will be common to find morality, intellectuality and devotion to the higher principles of life the marked features of the natal chart. Humanity is to-day building the future environments of this our Earth planet, and under these conditions, as humanity realizes its responsibilities to the coming generations and realizes its own power, that is, the power of creation from the thought plane, the thinking individual will be more careful of his thinking and thus become master, for, according to this divine science, we become that which we think, thus realizing that fate is not arbitrary but is acquired.

Thus as the ego goes forth from Universal Spirit unindividualized, it is enchanted by the parental influence and is forced into a new expression of life at the time of such configurations of planets and signs as take it into certain courses of action as may incline it strongly thereto. All the time the soul in its so-called suffering is still free and is not left without a remedy, for it can, if it choose, set the will against the stellar ruling; in the primary efforts it may cause itself much suffering, but it will obtain a new set of experiences to the current of its past, thus unbinding its own chains and gradually altering habits that may incline the soul in its first expression to individual consciousness. This may be termed ruling the stars or ruling one's own nature by one's free will. Thus, for example, suppose that in your own natal chart you came into mortal expression at a time when the Moon was placed in the eighth house in the sign of Virgo, and this was in square aspect to Mars placed in the sign Sagittarius in the eleventh house. The passions are strong, the mind is more concerned with the objective, though in this case the elevation of Jupiter in the tenth house in sextile to the Moon tends to mitigate the influence, Jupiter being also in trine to the Sun, which planet has strong rule in the individuality and will of the native. However, without this mitigating influence you can realize something of what the suffering would be from this one aspect, for the mind would have dominated the senses only after a long struggle, though finally the will would be the master and you would have emerged from the struggle victorious, having transmuted the square of sorrow into an aspect of joy. Then another aspect requires attention. Take the planet Mercury entering the sign Aries in third, square to Saturn in Cancer in the sixth house, which aspect mould cause much morry and anxiety of mind regarding the affairs of the first house, Aries, third house and the fourth house Cancer and the sixth house. Even though Saturn was receiving the trine of Jupiter from the tenth and the same aspect from the Sun in the second house, you can realize the power of this square aspect in the physical expression and can well understand how the soul would be hampered which was burdened with this aspect and receiving no mitigating influences to assist in alleviating it. Another instance is the square aspect of Vnus to Saturn. Venus being in the same sign as Mercury and the third house. True, your own will would not affect the thoughts and environments of others with as much power as your own. Thus it would be necessary for the active expression of the free will to so build up the environment that the so-called afflictions would not manifest with the degree of intensity as before. Thoughts build character, and environment comes from action. Aspiration, desire for the highest becomes capacity. Repeated thoughts become tendencies; will to perform becomes action; experience becomes wisdom; painful experience becomes conscience.

In further considering the nativity in its directional parts we must take the mundane parallels, both direct and converse, then follow the rapt parallels. We have heretofore given an interpretation of the mundane parallels, their equivalent distances, etc. The direct directions must be considered first.

Taking the chart upon which we have calculated heretofore, as the semi-arc of the Sun or Moon is to its mid-distance so is the semi-arc of planet moved to its second distance, from which, to find the arc, the primary from the mid-distance or nadir must be subtracted therefrom. For example, to direct the Moon to the parallel of Mars in the figure mentioned by direct direction in mundo, we proceed as follows: as the semi-arc of the Moon, that is, ninety-four degrees and twenty-five minutes in proportional logarithms equal nine and seventy-one and nine hundred seventy-eight ten-thousandths, as this amount is to the mid-distance of Moon, which we find to be thirty-one degrees and thirty-three minutes in proportional logarithms equal seventy-five and six hundred twenty-seven ten-thousandths, so is the semi-arc of Mars, which is found to be one hundred and thirteen degrees forty-one minutes equals in proportional logarithms nineteen nine hundred fifty-eight ten-thousandths, so is this to the secondary distance of Mars, which we find to be thirty-seven degrees and fifty-nine minutes, which equal in proportional logarithms sixty-seven five hundred sixty-three ten-thousandths. Therefore, taking the secondary distance of Mars, which is thirty-seven degrees fifty-nine minutes, from which is subtracted its primary distance from the mid-heaven, which is seven degrees and five minutes, we have a remainder of thirty degrees and fifty-four minutes, which sum constitutes the arc of direction of the Moon parallel Mars by direct direction in mundo. It will be observed that in this particular example Mars must be moved onward until it arrives beyond the cusp of the ninth house, where the aspect of Moon's parallel is completed; therefore the problem is really to first calculate the distance Mars must have traversed from the tenth to be the balance in mundane power or distance of the

Moon's radical mid-distance. The next step is to find the intercepted arc, which is at a point between these. This we may best term the secondary distance and the radical or primary positions of Mars, that is, its mid-distance, and we find the difference gives the arc of direction. After a little careful consideration I feel this will be quite clear to the student. Then we come to converse motions or directions. These have a slightly different formulae, though it is really only apparently so, for in these converse directions the luminary's semi-arc occupies the third term of proportion. In other words, to make more clear, the third term is the semi-arc of the moved or directed planet; it matters not whether these be direct or converse directions. The student is apt to become confused in these two directions, as it is the source from which error arises in the calculations of modern students of this science. However, I will illustrate clearly by example.

We will consider our subject of directions, and more particularly of converse directions, which apparent variance from the ordinary rules we have made mention heretofore, keeping in mind that the luminary's semi-arc ever occupies the third term of proportion. For example, we may take this illustration, the Sun to the parallel of the planet Saturn converse direction in mundo. Thus, as the semi-arc of Saturn, which we find in the figure first mentioned in this subject of directions to be one hundred and twenty degrees and five minutes, the proportional logarithms to equal nine and eighty-two four hundred twenty-one ten-thousandths, as this amount is to the mid-distance arc of Saturn, which we find to be seventy-four degrees and sixteen minutes, equals in proportional logarithms thirty-eight and four hundred four ten-thousandths, so is the semi-arc of the Sun as the third term of proportion, that is, seventy- two degrees and fifty-four minutes equals in proportional logarithms thirty-nine and two hundred fifty-four ten-thousandths so is the semi-arc of the Sun to the secondary distance of the Sun, that is, forty-five degrees and five minutes, equals in proportional logarithms sixty and one hundred twenty-three ten-thousandths. Then from the primary distance of the Sun from the nadir or cusp of the fourth, which is sixty-one degrees and thirty-two minutes, subtract the secondary distance of the Sun, which we have above to be above forty-five degrees and five minutes. As a remainder we have sixteen degrees and twenty-seven minutes, which equals the arc of direction of the Sun in parallel of the planet Saturn converse in mundo. The student will observe in this example (this being a converse direction) the greater luminary is moved onward until it stands in exactly the same relation to the lower mid-heaven or nadir, from which calculation is made for the reason that it is formed below the Earth as the planet Saturn radically posited does.

Then, too, the same results might be obtained by computing from the upper meridian, though the student would find no gain in this and only a more complex and cumbersome method of arriving at the final results. The student will observe that when a parallel will also be the conjunction or opposition of the bodies, it need not be calculated, for the arc of direction will ever be the same as the aspects heretofore mentioned and presumed to have been already obtained. Then, considering the rapt parallels: In the consideration of these, the radical places of the planets are not taken into account with reference to the completed aspect, as in the parallels heretofore dealt with, for the reason that their places are supposed to be carried on by the rapt motion of the Earth, or, according to students of ancient times, by what they termed the *primum mobile,* in such a manner that when a certain number of degrees of right ascension has passed over the meridian two planets will arrive at, should they be harmoniously posited at the moment of birth, in order to justify the event within the ordinary lifetime at equal distances from the meridian angles, and thus these differ from the foregoing mundane parallels, inasmuch as while in the latter the planet directed to remains immovable in the horary circle of position, the directed planet bearing eventually a relationship thereto.

We propose to throw light upon the great science which various individuals all down through the history of the world have sought to unravel its mysteries with more or less accuracy. Simplified forms of working out prognostications from aspects, and above all accuracy, are absolutely necessary to a perfect exposition of the science, and it is our purpose to give this sufficiently clear so that it will be as demonstrable and as accurate as the science of mathematics, which was really taken from astrological symbology, as we have shown heretofore.

Now, to continue our subject of rapt parallels. In all these terms it will be well to refer first of all to the Dictionary of Astrology which gives interpretations of these terms. In mundane parallels we find the calculations are made from the radical places of planets as though such planet were immovable in the circle of position, while the planet that is being directed to it in time bears relationship to it, that is, comes into aspect with it, while in the rapt parallels the places of both are carried forward and from thenceforth in their computations really bear no direct proportion to the radical places in the figure.

We may give a rule here that can be applied, simple and accurate in form; that is, first add together the semi-arcs of planets for obtaining the first term, then take the difference of the right ascension for the third term, while for the second term use the semi-arc óf the directed planet, which will ever be the planet that is approaching, or applying to the zenith or the nadir when the aspect is complete, and the results obtained will be the secondary distance of the directed planet. Then in obtaining the arc of direction, it will only be necessary to find the difference between the secondary distance of the planet directed and the primary distance of this planet. It will be well to give an example here of this calculation, using the same figure as given heretofore. For instance, we will direct the Moon to the rapt parallel of the planet Mars. First, we find the semi-arc of the Moon is ninety-four degrees and twenty-five minutes and the semi-arc of Mars is one hundred and thirteen degrees and forty-one minutes, which, added together, we find the sum equals two hundred and eight degrees and six minutes. We then obtain one-half of the total of semi-arcs, dividing by two, which gives one hundred and four degrees and three minutes. This is the first term. In order to obtain the third term, we find the difference of the right ascensions of planets involved in this wise. The right ascension of the Moon we find to be one hundred and eighty-one degrees and twenty-three minutes. The right ascension of Mars is one hundred and forty-two degrees and forty-five minutes. The right ascension of the Moon being greater and that of the planet Mars less, we then subtract the right ascension of the planet Mars from the right ascension of the Moon. This gives a remainder of thirty-eight degrees and thirty-eight minutes. Then to find one-half of the remainder, from subtraction of right ascensions divide by two, which gives nineteen degrees and nineteen minutes, which sum we may take as the third term in calculation.

Rapt parallels must be given special attention by the student as well as working out the converse directions. Both are important in obtaining accurate results. In our last meeting we have obtained the first term in the example by adding together the semi-arcs of planets involved, also the third term by obtaining the differences of the right ascensions. We now proceed and the semi-arc of the directed planet, that is, the Moon, will be used as the second term in the problem. Thus we have a product of one hundred and four degrees and three minutes of the adding of semi-arcs and one-half of their total, which equal in proportional logarithms nine and seventy-six and one hundred and ninety-seven ten-thousandths, as this sum is to one-half the sum of the Moon's semi-arc, which is forty-seven degrees and twelve minutes, which in proportional logarithms equal fifty-eight one hundred thirty-three ten-thousandths, so is one-half the difference of the right ascensions of the Moon and Mars, which we have found to be nineteen degrees and nineteen minutes, which equal in proportional logarithms ninety-six nine hundred and thirty-four ten-thousandths. Thus we have the problem in this wise: as one-half of the sum of semi-arc of Moon and Mars is to one-half the sum of the Moon's semi-arc, so is one-half the difference of the right ascension to one-half of the secondary distance of the Moon which we find to be eight degrees and forty-six minutes, which equals in proportional logarithms one and thirty-one two hundred sixty-four ten-thousandths.

Now it will be necessary to learn the whole secondary distance of the Moon. This we find by multiplying one-half Moon's secondary distance by two which gives Moon's secondary distance as seventeen degrees and thirty-two minutes. Thus we have the primary distance of the Moon as thirty-one degrees and thirty-three minutes and the secondary distance of the Moon as seventeen degrees thirty-two minutes, and by substracting the Moon's secondary distance from its primary distance we have as a remainder fourteen degrees and one minute which is the arc of direction of the Moon in rapt parallel to Mars. The student will observe that as these problems provide large numbers it is better and more simplified to work out one-half or even less of the amounts involved, keeping in mind to increase the proportional part of secondary distances when obtained by so many aliquot parts as have been worked with, and in some instances the two bodies will be located in opposite hemispheres and in such instances the opposite place of the one which will be receding from the mid-heaven or nadir on formation must be taken, for in these directions the arcs must be of the same denominations. This procedure will only necessitate one hundred and eighty degrees being added to or subtracted from its right ascension as the occasion will demand, the semi-arc remaining as before. The difficulty will lessen under application and the more application the quicker and more accurate will be the results obtained.

END OF VOL. VI.

Alvidas was really Henry Clay Hodges, who worked as
an astrologer in Detroit, MI, at the turn of the century.